Understanding Science

In a time when the role of science in society is under threat, this book provides a timely and accessible text that can be used to learn or teach both the theory and practices of science, and how they are interconnected. Chapters 1–7 introduce the major approaches to the philosophy of science using simple language and examples that are easy to understand. Chapters 8–10 build on the philosophy of science to explain science practices such as publication, bibliometrics, experiments, the use of statistics, research ethics and the academic career. This book emphasizes how and why science is the most reliable source of knowledge and how society is dependent on science to make informed decisions. It primarily targets science students but is also accessible to general readers interested in understanding how science works. It is ideal as a textbook for intermediate to advanced students majoring in any science (or engineering) subject.

NIKLAS JANZ is a professor in evolutionary insect ecology at Stockholm University, Sweden. His research has mainly focussed on the evolutionary association between insects and plants. He has taught extensively on ecology, evolutionary biology and the philosophy of science. He has also published two novels for young adults.

SÖREN NYLIN is a professor of animal ecology at Stockholm University, Sweden, where he also teaches theoretical and practical aspects of science. He has published a large number of scientific articles, primarily on insect evolutionary ecology, and is an elected member of the Royal Swedish Academy of Sciences.

'This book offers an intellectual journey well worth taking for students and established researchers alike. The authors take a unique, historical approach to exploring the nature of science that will be useful for those who want to ground their communication and engagement activities in a nuanced understanding of the philosophical tradition. The everyday examples and clearly parsed definitions make it eminently readable and an important contribution to the curriculum, as well as a timely, valuable resource for affirming why science is worthy of trust.'

Jamie Bell, Association of Science and Technology Centers, USA

'Science needs this book. With the critical role of science in human advancement under threat, we need all world citizens to understand what science is, what science does, and how science works. But before that can happen, scientists – from beginning students to established researchers – must think about how these issues shape their work and how they can (and must) communicate them to broader society. *Understanding Science* provides just that perspective, ranging from philosophy to ethics to the logistics of funding and publication, in an accessible and thought-provoking way.'

Stephen Heard, University of New Brunswick, Canada

Understanding Science
The Web of Trust

NIKLAS JANZ
Stockholm University

SÖREN NYLIN
Stockholm University

Shaftesbury Road, Cambridge CB2 8EA, United Kingdom

One Liberty Plaza, 20th Floor, New York, NY 10006, USA

477 Williamstown Road, Port Melbourne, VIC 3207, Australia

314–321, 3rd Floor, Plot 3, Splendor Forum, Jasola District Centre, New Delhi – 110025, India

103 Penang Road, #05–06/07, Visioncrest Commercial, Singapore 238467

Cambridge University Press is part of Cambridge University Press & Assessment, a department of the University of Cambridge.

We share the University's mission to contribute to society through the pursuit of education, learning and research at the highest international levels of excellence.

www.cambridge.org
Information on this title: www.cambridge.org/9781009672849

DOI: 10.1017/9781009672863

© Niklas Janz and Sören Nylin 2026

This publication is in copyright. Subject to statutory exception and to the provisions of relevant collective licensing agreements, no reproduction of any part may take place without the written permission of Cambridge University Press & Assessment.

When citing this work, please include a reference to the DOI 10.1017/9781009672863

First published 2026

Cover image: Photography by Lana Galina/Moment/Getty Images

A catalogue record for this publication is available from the British Library

A Cataloging-in-Publication data record for this book is available from the Library of Congress

ISBN 978-1-009-67289-4 Hardback
ISBN 978-1-009-67284-9 Paperback

Additional resources for this publication at www.cambridge.org/janznylin.

Cambridge University Press & Assessment has no responsibility for the persistence or accuracy of URLs for external or third-party internet websites referred to in this publication and does not guarantee that any content on such websites is, or will remain, accurate or appropriate.

For EU product safety concerns, contact us at Calle de José Abascal, 56, 1°, 28003 Madrid, Spain, or email eugpsr@cambridge.org

Contents

	Preface	page ix
1	**What's So Special about Science?**	1
	1.1 'I Could Be Wrong'	1
	1.2 Is There Anything Out There?	2
	1.3 Making the Subjective Objective	4
	1.4 Facts and Theory	5
	1.5 'Follow the Science'	9
	1.6 Understanding Science	10
2	**Observation-Driven Science**	12
	2.1 Mary's Butterflies	12
	2.2 Early Empiricism: Francis Bacon	13
	2.2.1 The Baconian Method	16
	2.2.2 Did Scientists Historically Follow the Baconian Method?	17
	2.2.3 Bacon's View of Progress	18
	2.3 Later Empiricism: The Positivists	18
	2.4 The Limitations of Empiricism	19
	2.4.1 Logical Problems	20
	2.4.2 Observations Are Fallible	21
	2.4.3 Observing without Theory Is Ineffective	25
	2.5 The Place of Empiricism in Science	25
3	**Hypothesis-Driven Science: Falsificationism**	28
	3.1 Mary's Butterflies	28
	3.2 Karl Popper and Falsificationism	30
	3.2.1 The Logic of Scientific Discovery	31
	3.2.2 A Shift in Perspective	31

	3.2.3 Falsifiability	34
	3.2.4 Clarity and Precision	35
	3.2.5 Predictions	36
3.3	Mechanisms and Change	37
3.4	Ad Hoc Hypotheses	38

4 Hypothesis-Driven Science: Limitations and Alternatives — 41
4.1 The Limitations of Hypothesis Testing — 41
    4.1.1 The Duhem–Quine Problem — 42
    4.1.2 Falsification and Confirmation — 43
    4.1.3 Scientific Progress — 44
    4.1.4 The Logic of Scientific Discovery? — 45
4.2 The Place of Hypothesis Testing in Science — 46
    4.2.1 Taking a Step Back — 47
    4.2.2 Bayesianism — 48
    4.2.3 Bayesianism as a Philosophy of Science — 51

5 Paradigm-Driven Science — 55
5.1 Mary's Butterflies — 55
5.2 The Philosophy of Paradigms: Thomas Kuhn — 56
    5.2.1 Normal Science and Revolutions — 56
    5.2.2 Incommensurability — 58
5.3 The Limitations of Paradigms — 59
    5.3.1 Where Are the Revolutions? — 59
    5.3.2 The Preservation of Knowledge — 60
5.4 The Place of Paradigms in Science — 62
    5.4.1 Research Programmes — 63
    5.4.2 The New Experimentalism — 65

6 Science as a Social Activity — 67
6.1 Mary's Butterflies — 67
6.2 Public Knowledge — 68
    6.2.1 The Problem of Demarcation — 69
    6.2.2 The Growth of Science — 71
    6.2.3 Scientific Institutions — 72
6.3 A Social Definition of Science — 75
    6.3.1 The Search for Consensus — 75
    6.3.2 The Social Definition and Philosophy of Science — 77

7 Synthesis — 79
7.1 What Do You Mean by 'Science'? — 79
7.2 Search for Consensus as a Line of Demarcation — 81

	7.3 Being Scientific	85
	7.3.1 The Marks of Good Science	86
	7.3.2 When Science Is Not Scientific	88
	7.3.3 The Marks of 'Non-Science'	88
	7.4 The Web of Trust	89
8	**Science in Practice: Publishing**	**91**
	8.1 The Publication Process	91
	8.1.1 The Format of Scientific Publications	92
	8.1.2 How to Read Papers Effectively	96
	8.1.3 Publishing	97
	8.1.4 Finding Publications	103
	8.2 Citations and Impact Factors: Bibliometrics	104
	8.2.1 Author Citation Metrics	105
	8.2.2 Journal Citation Metrics	106
	8.2.3 University Rankings and Research Evaluations	108
9	**Science in Practice: Data**	**110**
	9.1 Scientific Data in the Light of Philosophy	110
	9.1.1 Data from Pure Observations	110
	9.1.2 Data from Structured Observations	111
	9.1.3 Data from Experiments	113
	9.1.4 Models in Science	115
	9.2 Dealing with Variation	116
	9.2.1 Variation and Statistics	117
	9.2.2 Falsification and the Null Hypothesis	119
	9.2.3 Significance Thresholds	121
	9.2.4 Predicted Results and Chance Findings	123
	9.2.5 Alternative Hypotheses and Bayesian Statistics	125
	9.3 Reviews and Meta-Analyses	126
	9.4 Combining Evidence: An Example	127
10	**Science in Practice: Academia**	**130**
	10.1 Academia and the Competent Researcher	130
	10.1.1 The PhD Education	130
	10.1.2 The Academic Career	132
	10.1.3 Equal Opportunities in Science	133
	10.1.4 The Invisible College	135
	10.2 Academic Freedom	137
	10.3 Funding	138
	10.4 Ethics in Science	139

		10.4.1 Research Subjects	140
		10.4.2 Human and Animal Research	140
		10.4.3 Plagiarising	140
		10.4.4 Publication Misconduct	141
		10.4.5 Conflicts of Interest	141
		10.4.6 Cheating and Manipulation of Results	142
	10.5	Science Outside of the Academy	143
		10.5.1 The Challenge of the Consensus	143
		10.5.2 Evaluating the Current Scientific Consensus	144
		10.5.3 Making Science Actionable	146
11	**Epilogue**		149
	References		153
	Index		156

Preface

This book was born out of the experiences of two decades of joint teaching of philosophy of science and 'science in practice' to undergraduate and graduate students in biology at Stockholm University. It is primarily intended for science students and early career scientists, whose main interest is doing or communicating science and not philosophy as such, because we believe that any person working in or with science should be able to articulate what science is and what makes it special as a means of gaining knowledge about the world. This includes having some appreciation of the main attempts by philosophers to tackle this question, as well as of how these philosophies influence how science works today. Online resources can be found at www.cambridge.org/janznylin for teachers and supervisors interested in using this book as a textbook. With this said, as science has a profound impact on society, a basic understanding of what makes it work is arguably of importance for every citizen. Hence, we have tried to keep the text straightforward and not too philosophically esoteric, in the hope that the text should be accessible also for readers outside of the academic system who are curious about how it works.

We are biologists, and both of us have spent our careers investigating the evolutionary ecology of butterflies and their relationship with the plants they feed on. This book will naturally be coloured by this background, which gives it a somewhat different vantage point than most books about philosophy of science. Much of the foundational work about general philosophy of science has been written from the point of view of physics. While there is much in common between all scientific disciplines, there are also some noteworthy differences, and we hope that our deviating background may provide some added perspective. Still, while we do draw from our experience in evolutionary biology, and the unique challenges that are associated with biology, our aim remains to reach a general understanding of the nature of science itself. Hence, you will notice that this book contains quite a few examples from biology (and butterflies!), but they invariably serve to illustrate general points and should require no background in biology to understand.

We have several goals with this book. One is to give a quick overview of the historical developments in philosophy of science, by condensing them to the most influential approaches. With this we aim to reach an understanding of what science really is, and why this has been so difficult to grasp over the years. This is the main focus of Chapters 1–7. Chapters 8–10 instead focus on science in practice, and how science works and is used inside and outside the academy. The way science is done is a moving target. Technological and societal developments will continue to challenge and change some of the current practices, perhaps fundamentally. Still, it is worth looking at the way science is currently done to highlight which fundamental functions need to be fulfilled. It is of course important for a working scientist to understand why we do science the way we do, but science today is also immensely influential in our daily lives. Perhaps more than it has ever been. This has created an increased need for everyone – not only the active scientist – to have some fundamental understanding of how science works, and how it can be used to aid societal development.

At the same time, there has been a rise in the deliberate spread of science-scepticism and misinformation, muddying the waters and making it increasingly difficult to find and propagate accurate information. In our own teaching, we have seen an increased need to go from only teaching students the importance of being 'critical', towards also emphasising the role of trust. In a sense, it is a misunderstanding of what it means to be critical that lies behind the spread of misinformation. It is not difficult in itself to be critical, if that entails just rejecting information. The difficult part is to evaluate information and decide what can be trusted and what cannot. How do we know what we can trust? Why is it that we consider 'scientific' findings superior to other types of information? How can we justify this special position we have given science? We have chosen to give this book the subtitle *The Web of Trust* to highlight the crucial role that trust plays in science and the various ways that such trust is earned and assured. We hope to convey a multifaceted but still coherent picture of science, how and why it works, and what role it plays in our lives, inside and outside academia. We will piece together the various components that comprise what we call the web of trust, and we hope to show why this web is so crucial to maintain and defend.

1
What's So Special about Science?

The fallibility of knowledge and the web of trust.

1.1 'I Could Be Wrong'

How do you know that you know something? Throughout history, people have attempted some very different approaches to this question, based on, for example, authority, observation, reason, consensus, introspection, or religious revelation. We think it is safe to say that one of the many attempted approaches has been singularly successful. It has allowed humankind to develop knowledge even about phenomena beyond our direct perception, such as molecular and sub-atomic processes, as well as deep space, and fuelled practical applications such as medicine, aviation and electronics. The approach we refer to is of course science. But what precisely do we mean by 'science'? The answer to this question is surprisingly elusive. When pressed about it, could you articulate what you mean by science? Does it matter?

There is a prevailing sentiment among scientists and philosophers alike that both groups of people should stick to what they do best. That is, scientists should stay in their labs and leave the philosophy of science to the professionals. It may be true that all serious work requires specialisation, but this does not mean that philosophy of science is of no consequence to the scientist. It is important to have at least a basic understanding of different approaches to how we build scientific knowledge, their merits and potential problems, and 'how we know that we know something', if for no other reason than to understand why we do science as we do. We are also convinced that any working scientist – or really, anyone using scientific results – should be able to articulate what makes science different from other means of gaining knowledge. You can sometimes hear that science doesn't have to be properly defined, since 'we know it when we see it'. This may be true to some extent, but it is not very helpful when trying to explain or defend the special position that science has in our society. In today's world of fast-spreading misinformation and widespread science-scepticism, this has become even more important. There is an increasing need to explain why we should believe statements of scientific consensus over any other kind of opinion. How can we defend the special standing of science in society if we cannot articulate what we mean by it?

Understanding the nature of science, and how scientific knowledge is built, is also important in order to understand its limitations. The chapters in this book should make it clear that to believe that science is the best available means of gaining knowledge about the world is perfectly compatible with the realisation that it is also fallible. The demonstration that a particular published scientific result is wrong does not mean that we cannot trust science. Quite the contrary. In one important sense, it is this internalised realisation that 'I could be wrong' that makes science so powerful. It is what allows it to progress by learning from its mistakes. Few other human endeavours have such error correction built into its very foundation. This is such an important aspect of science that we may call it a cardinal rule.

Understanding what makes science special, what it is that makes it *work*, also means that we learn what aspects of it that we need to protect. Science has been remarkably successful, but some of its basic tenets are fragile and may need to be guarded, be it from deliberate attempts to undermine the trust in science or well-meaning administrative changes.

1.2 Is There Anything Out There?

To start from the beginning, how can we know that there is anything out there to learn about at all? In other words: is there an objective reality that exists independently of you and your senses? This has been (and still is) a contested question among philosophers, especially with respect to phenomena that cannot directly be experienced through our senses. And even if such a reality exists, how can we know that our sensual experiences or our theories have any true connection to this objective reality? Philosophers have taken up different positions on these questions – commonly called realism and antirealism. In short, realists assume that we can learn about the true nature of the world by careful observation and testing of hypotheses. The fact that we can make increasingly precise predictions about the world, and manipulate it in complex and subtle ways, suggests that we are getting closer to a true understanding of it. Antirealists, however, argue that if there is such a thing as an external mind-independent reality, we can never be sure that our theories have a true correspondence to it. Instead, they view scientific theories as pragmatic constructs that allow us to make predictions, but that are not necessarily accurate descriptions of an underlying reality.

This question may be interesting from a philosophical standpoint (and perhaps impossible to truly resolve), but we believe that in practice the different positions tend to converge, at least for working scientists. Few realist scientists would argue that we can say with certainty that our observations and theories are true representations of the world. If they did, they would violate the cardinal rule we just mentioned, that 'I could be wrong'. They may see them as tentative truths but must be willing to accept that new observations or theoretical developments will require them to change their interpretation. In a similar way, few antirealist scientists would argue that trying to learn about what is beyond our own experience is just a mental exercise, since that would make most scientific work meaningless. Even if an antirealist believes that we can never know for sure if our concepts actually correctly describe the 'true' reality, trying to build increasingly sophisticated theories with better predictive power will still be an important goal.

Thus, in the end, it seems to us that more qualified versions of realism and antirealism end up being similar in scientific practice. The antirealists just emphasise the inherent uncertainty of knowledge a little more. Moreover, we would argue that few scientists really spend much time worrying over whether our theories actually bring us closer to an accurate representation of reality (or to what extent we can say that such a mind-independent reality even exists). Most of us are content to assume that either the world exists or it behaves as if it does, and for all practical purposes that is the same for us. Nevertheless, this issue has been important for many formulations of philosophy of science, and we may have reason to briefly revisit it. The important thing to note is that doing science involves making inferences in the face of uncertainty. Almost by definition, a scientific study attempts to answer a question that has not yet been asked. There is no set answer you can look up at the end of the study to check if you got it right. Scientific conclusions are almost always reached by conjecture, and this is true wherever you place yourself on the realist–antirealist scale.

However, there is a related – and perhaps more practical – issue with our perception of reality, and how we can know anything about it. We are constantly made aware of events and phenomena in the world, not just new scientific discoveries but all kinds of events happening worldwide. Almost all of this is beyond our immediate ability of personal verification. Strictly speaking, there is not much of what we call reality that is possible for us to confirm through personal experience. There may have been a time when most of the events that came to the awareness of a typical person concerned their immediate local reality. If so, this is certainly not the case today, when most of the events that we are made aware of, for example, through news and social media, are not part of our own personal reality. How can you be *sure* that the events that are reported in the news have even taken place? In most cases, you simply cannot. Almost all information that we encounter requires a certain degree of trust for us to accept. But who and what can you trust? It is not easy to evaluate the various sources of information that we are exposed to, and conspiracy theories exploit this need for trust, by sowing uncertainty and making you question what you would normally rely on. *Why* should you trust the news rather than the conspiracies on the internet? Why indeed should you trust science?

After all, trust is just as important when it comes to scientific information as it is for everyday knowledge about what is happening in the world. Indeed, science is highly *dependent* on trust. The realist–antirealist debate concerns itself with whether it is possible to attain true knowledge about the world, but in order to do science in a meaningful way we also have to trust other people's observations and conclusions, at least to some extent (as we shall see in later chapters, especially Chapters 6–8, there is an important role of *dissent* among scientists as well, as a quality-check and to promote discussion and growth of knowledge[1]). Much of what we do as scientists builds directly on what others claim to have found, and without some trust it would be impossible to build on such knowledge. Moreover, the technical nature of scientific papers makes them difficult to penetrate for anyone not in the field, and relatively few people are truly able to assess and verify any particular scientific claim.

Thus, to understand science, it is of fundamental importance to ask why we should trust a paper in a scientific journal more than any other source of information, such as an authoritative social media posting. It turns out that science does have some built-in mechanisms to build trust, and we will get back to these in Chapters 8–10. For now, we will begin by turning the question around and look at it from the reverse perspective. How do you make your own subjective observations worthy of other people's trust? Is there anything you can do that does not amount to authoritative posturing?

1.3 Making the Subjective Objective

Since we are all different individuals, without direct access to the experiences of others, all our observations are inherently subjective. The natural sciences, however, are typically not interested in subjective knowledge as such, but

---

[1] As Jacob Bronowski once put it (in Science and Human Values, 1956), science is based on these two pillars: trust and dissent, the latter in the form of constructive scepticism.

strive for objective knowledge about the world. How is that even possible when our observations are fundamentally subjective?

If you observe something, say an oak tree at a distance, how can you be sure that you are indeed observing an oak tree? Perhaps you are looking at a cleverly placed photograph, or what you see is actually a reflection in a mirror? Since we have two eyes that allow stereoscopic vision, the first thing you would do to make sure you are observing a large tree at a distance, rather than a small image of a tree close up, is probably to move your head or your point of view. This would provide you with enough useful information to evaluate if it is a real tree you see. In other cases, we could also involve other senses or make the observation under different conditions. Much of those actions are done without thinking, but they are nevertheless active. We *do* things to establish what it is we see. Observation is an active process, not a passive reception of information.

You may now have convinced yourself that what you think you observe is correct. But this conviction is still subjective, and what if the tree is just an illusion, a trick of your mind? Again, there are things you can do to make your subjective observation more objective. For example, you can ask someone else to make the same observation, and hopefully they agree that they are observing an oak tree. In this simple case, this may be enough to convince you and your fellow observer, but scientific observations are often made at the edge of current knowledge, and understanding what such observations mean is by no means easy. Making subjective observations more objective is an important part of the scientific process and is not as simple as it may intuitively seem. There are a number of steps that researchers typically take to increase the objectivity of observations. These steps include careful method descriptions to increase repeatability of the observations, transparent calculations and data transformations, calibration of instruments, 'blinded' observations and so forth. Or, like we just did with the oak tree, someone else can repeat the observation (or experiment). Scientists typically go to great lengths to convince their peers that thier interpretations are accurate, and all of these efforts involve shifting the observation from the subjective towards the objective.

1.4 Facts and Theory

One of the most commonly held notions of science, at least among laypeople, is that it is 'based on facts'. But what does that mean? What is a fact? A typical answer would be an *unbiased observation* – an observation that is not tainted by assumptions or theory. This is, however, a notion that does not hold up to scrutiny.

If we return to our observation of the oak tree, it seems straightforward enough to approach the kind of unbiased observation that would make it an unproblematic fact. But even if you have established that it is not an illusion or a photograph, how do you know that you are looking at an *oak tree*? What, indeed, is an oak tree? The very words 'oak' and 'tree' are actually not really factual statements but are best thought of as (well-established) theories. Within botany, a tree is a perennial plant with an elongated woody structure that supports its external branches, leaves and flowers. It is not a taxonomic group but a growth form that has evolved independently in many distantly related groups of plants. As with many such concepts, its boundaries are not clear-cut. In some instances, it may be difficult to decide whether you are observing a small tree or a large shrub, and there is disagreement about whether a palm tree should be regarded as a tree (because it has no secondary growth). Presumably, you would also not call a young oak sapling a 'tree', even if it may well become one. It should be clear from this that 'observing a tree' actually requires some prior knowledge of what a tree is. This is even more evident when you claim that it is not only a tree but an oak tree. An 'oak tree' is an even more elaborate theory, one that involves the concept of biological species and the phylogenetic relationships between them. In this case, an 'oak tree' refers to a plant in the genus *Quercus*. The genus *Quercus* is in fact a large group of about 500 related species, and depending on your prior knowledge, and where you are, the statement 'I see an oak tree' could mean some different things. In northern Europe, where the authors of this book live, there is only one species of oak growing in the wild, so if this is where you made the statement, this prior knowledge would imply that you are indeed observing a tree of the species *Quercus robur*. If you made it somewhere else, perhaps in southern Europe or the Americas, the statement would be less specific, since there are several species of *Quercus* in these regions, some quite different from each other.

A consequence of this is that when we say that the oak tree is an observable fact, this 'fact' cannot be said to truly refer to the actual object itself. We have already established that we do not have direct access to the external reality, only our subjective interpretation of it. So strictly speaking, when we say 'I see an oak tree', it is a *statement* about the object we observe – a theory if you like. This matters, because as our statements can (at best) only be tentative representations of a true reality, all 'facts' must be regarded as tentative theories. Indeed, the history of science is full of observations of phenomena that are now considered dubious or even non-existing.

For example, for a long time, scientists made observations confirming the existence of phenomena such as the ether and phlogiston. These concepts originated in ancient Greece but were used to explain several natural phenomena well into the nineteenth century. In classical Aristotelian physics, the cosmos was divided into the sub- and super-lunar spheres. The space between the earth and the moon consisted of the four elements: air, earth, water and fire. Above the moon was the unchanging sphere made up of the fifth element, ether. Closer to our time, the ether was used to explain the movement of light through a vacuum, and Newton used it to explain gravity. The existence of the ether was thus repeatedly confirmed by observations such as light travelling from stars and bodies falling towards the earth, and it was not until Einstein formulated his theory of relativity that the ether finally fell into the scientific bin. Phlogiston was thought to be a substance released during combustion, and the loss of weight when substances such as wood burned was attributed to the release of phlogiston. As phlogiston was considered to be lighter than air, its existence also provided a neat explanation for why flames move upwards, away from the gravitational pull. Again, observing how burning objects lost weight and how flames moved upwards confirmed the existence of phlogiston. We can still make the exact same observations today, but now we interpret them differently.

A more recent 'discovery' was made by the accomplished French scientist René Blondlot in 1903. Blondlot had been experimenting with the newly discovered X-rays to determine if they are waves or streams of particles. He correctly determined that they are in fact waves, but during his further experimentation he made unexpected observations that he interpreted as evidence of something entirely new: a form of radiation which he called N-rays [1] (Figure 1.1). This caused quite a sensation at the time and many tried – usually unsuccessfully – to replicate Blondlot's findings, which among other things could only be done (according to Blondlot) if you didn't watch the phenomena caused by N-rays straight on; you had to use your peripheral vision. This may sound ridiculous in retrospect, but remember that phenomena that are invisible in themselves always require a certain amount of faith in the supporting axioms and in the methodology used to observe their effects. Blondlot, however, even registered the effects of N-rays photographically (Figure 1.1) in order to have what he deemed to be objective evidence. Scientists often choose to believe in phenomena that fit into their current models or whose existence provide neat explanations for other phenomena that would otherwise be hard to explain. To actually demonstrate the veracity of such a phenomenon is not trivial. In 1904, Robert W. Wood convincingly showed that N-rays in fact do not exist but were a product of Blondlot's incorrect interpretation of observations.

These examples should be enough to illustrate that the observations we make are to a large extent dependent on currently accepted theory. Even the

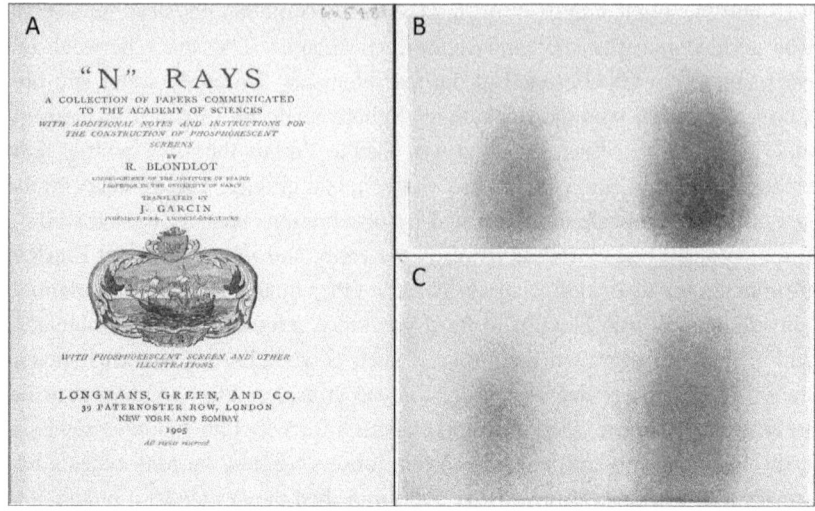

Figure 1.1 (a) The cover of a collection of papers describing the results of René Blondlot's experiments on the N-rays, in detail. (b) Photographic registering of the action of a small electrical spark without and with N-rays 'emitting from a Nernst lamp'. (c) Without and with N-rays 'produced by two large files'. N-rays were soon plausibly demonstrated to be non-existent, meaning that what was observed and 'objectively' registered by Blondlot must have been spurious phenomena. Source: Internet Archive.

simplest observation statements, such as 'I see a tree' require prior knowledge of the observed and an interpretive framework. This is not at all a trivial point to make. Students in biology often have to learn a large number of plants and animals by heart. One perhaps unexpected consequence is that after this exercise, words such as 'trees' or 'grass' will have taken on new meanings. Where the untrained eye would just see 'grass', the trained botany student will see several different species, all with their own distinct characteristics. Their eyes have access to exactly the same information, but in a very important sense they will actually see different things.

The types of observations that are made in scientific studies are typically even more theory-dependent. Very often, such observations are not direct but come in the form of instrument readings, and the construction and calibration of such instruments are heavily dependent on theory. Even when observing through a microscope with our own eyes, we rely on the optical theory of diffraction, reflection and refraction, among other things. Not to mention the highly sophisticated theory behind electron or laser microscopy. Clearly, observations are not as straightforward and unbiased as they may seem. We will explore such theory dependence further in Chapter 2.

A final complication worth mentioning before we leave this topic is that the terms 'fact' and 'theory' are used quite differently among scientists than outside academia. While a layperson may say 'that is just a theory' to insinuate that a statement has no factual basis at all, for a scientist a theory is a well-supported statement about the world. Indeed, you will rarely see or hear a scientist use the term 'fact'. There is a good reason for this, since it would run in the face of the scientific built-in error-checking mechanism. As stated at the beginning of this chapter, it is a basic tenet in science that we cannot say anything with absolute certainty, so using words that signal such conclusiveness feels out of place. Scientists further distinguish between *theory*, *hypothesis* and *prediction*. There is some overlap between these terms, especially in everyday language, but they do have distinct meanings. A theory is a well-supported set of cohesive ideas. A hypothesis is derived from a theory and is akin to a tentative truth or a qualified guess. Finally, a prediction is derived from the hypothesis and is a specific statement about what to expect in a given test situation – such as an experiment – if the hypothesis is true. For all intents and purposes, what lay people refer to as a 'fact' would for a scientist translate as a 'well-supported theory'.

1.5 'Follow the Science'

Not everybody accepts the superiority of science as a way to gain knowledge about the world. In fact, science-scepticism has gained much ground in recent years, following the rise of social media. These platforms thrive on engagement, which seems to encourage increasingly polarised opinions and ideology. The resulting factions can easily develop world views that clash with established views, and if they do, they can promote their own authorities, and dismiss the scientific experts. Resolute science sceptics probably cannot be reached with a book such as this one, but we do hope that it can address the problem indirectly by giving the readers a deeper understanding of science and hence some tools to explain and defend its standing as the most reliable source of knowledge.

You only need to browse any form of news or social media to realise that science has a special standing in society. References to science are commonly used to back up all forms of statements, including those that are really about politics, or even religion. 'Now even scientists say …'; 'scientific studies show …'; and similar sentences abound. During the COVID-19 pandemic, this became almost painfully clear. Thousands of scientific studies on COVID-related topics were performed during the pandemic, and to an unparalleled extent first drafts of such studies – not yet scrutinised and reviewed by scientific peers – were reported in the news. The results were then often immediately used to back up any opinion already held, and broadcasted in social media. The rally cry that we should 'follow the science' was routinely used by all sides of any discussion about what interventions should be used (or not) to mitigate the pandemic. This period offered people outside of the academic world unique insights into the inner workings of science, but it was also clear that in many cases, the scientific output was misrepresented due to a failure to understand the scientific process. In particular, it highlighted how difficult it is to embrace the fallibility of knowledge and the tentative nature of findings at the cutting edge of science. These features of the scientific process are easy to perceive as a weakness of science as a means of gaining knowledge about the world, but as we stated already in the beginning of this chapter, they are in fact integral to its success.

Remarkably, even science sceptics have a tendency to refer to 'scientific evidence' whenever it happens to suit their purposes. One example is the field of alternative medicine, such as homeopathy, where proponents often claim that it is not possible to test their methods using normal scientific practice. This, they say, is because every patient, every therapist–patient interaction and every situation is unique and not repeatable in the controlled way that science demands. Nevertheless, whenever a study of alternative medicine happens to obtain results that can be construed as support of such methods, proponents can still be quick to refer to them as evidence that their ideas work. Another example is climate change, where there is near-complete scientific consensus that the planet is warming and that the warming is largely caused by human activities. Non-believers will simultaneously dismiss these 'so-called experts' and use the few dissenters that exist to demonstrate that even scientists share their view. The fact that science sceptics also resort to using science to back up their claims when this is possible is a stark demonstration of the special standing that science has in our society as a source of trustworthy knowledge.

1.6 Understanding Science

Science-scepticism notwithstanding, it is hard to avoid the conclusion that science has been a remarkably successful human endeavour. Science has been an incredibly important part of the advancement of human societies during at least the most recent centuries, possibly much longer than that. With that in mind, it may seem somewhat paradoxical that the exact nature and definition of science is plagued by controversy and confusion. Indeed, the nature of science has been a topic of a heated debate that has been raging among philosophers for hundreds of years. Since most people appear to agree that there is something special about science, it is puzzling that it seems to be so hard to agree on exactly what we mean by it.

It is not that we lack attempts or suggestions to define and understand science, and, in this book, we will guide you through some of the most influential. We will also end by sharing our perspective on it and why we think that it has been so hard to reach a universal agreement. Before we present these ideas, we should point out that what activities people would typically want to include under the label *science* is not that coherent, which of course makes our task no easier. To some extent, this has cultural reasons. In some parts of the world, science is a rather inclusive term, typically referring to all subjects studied at universities, while in other places (such as the Anglo-Saxon world), science usually implies the natural sciences. Other subjects may need a prefix, such as the 'social sciences', or they are referred to as 'arts', or 'the humanities'. As individuals, we may hold different views on what qualifies as a 'real' science as well. It is worth keeping this in mind while reading this book, as the history of philosophy of science will be quite different in the natural sciences and the social sciences or humanities. This is not surprising, considering that it may not be a realistic goal to study human societies in the same way that you study electron particles.

Compared to the natural sciences, the research traditions in the social sciences and humanities are much more heterogeneous. Interesting as these traditions are, being biologists ourselves, our focus in this book will be on the natural sciences. We believe others are better suited to explain and interpret the research traditions of the social sciences and humanities. Nevertheless, we will briefly return to some of these research traditions in Chapter 7 to see how they fit into the view of science that emerges throughout its chapters.

But before we get to that point, we need to take a journey to see the various ways that philosophers have tried to tackle this question, how they relate to each other, and to what extent they succeed in their endeavours. Following this journey, we will take a closer look at the modern practice of science and its peculiarities. We will investigate the extent to which this practice seems to fit the theories of the philosophers but also highlight the components of what we have chosen to call the *web of trust*. What is it about science that makes it more trustworthy than other sources of knowledge? And how do you know which science to trust? And indeed, what does it mean to 'follow the science'?

2
Observation-Driven Science

Science is based on observations of facts, from which generalisations can be made.

2.1 Mary's Butterflies

We will use a story about Mary and her butterfly studies (and its continuation in Chapters 3 and 5–6) as a starting point for our illustrations of possible ways that science might be defined and described.

> Imagine, if you will, a woman calmly sitting on her porch, facing the somewhat unkempt backyard of the house that she shares with her husband and her eight-year-old son. It is a sunny morning in the early spring, and some snow still lingers in the darkest corners of the garden. The buds are just about to break on the birches, willows and currants that line the outskirts of the backyard. Spring flowers like crocus and daffodil are in bloom, scattered across the lawn in places where they have propagated themselves over the years, and some other herbs have also pushed their way out of the soil to show tender green leaves. Some of them grow where they are supposed to, others are weeds that don't obey such rules. The woman is smiling and seems pleased with the way the garden looks, weeds or not. Let's call her Mary.
>
> Mary sees something moving out of the corner of her eyes. A butterfly! Mary knows her butterflies, and there are not many species on the wing this early in the year, only the ones that brave the winter as fully-formed adults. This one has an overall darkish appearance as it flutters about in the garden. It is thus very probably a peacock butterfly, or *Aglais io* if we want to be more scientific about it (Figure 2.1). It is a very common species, and Mary sees it every year in her garden. The behaviour of this particular individual peaks her interest. It seems to give much attention to the weeds, flying slowly just above the greenery that is barely visible among the wilted grass from last year. Sometimes it lands briefly to check out a particular leaf, on occasion displaying the eyespots that confirm the identity of the species, but then sets off again. All of this suggests to Mary that this particular butterfly is a female looking for a suitable place to put her eggs.

2 Observation-Driven Science

Figure 2.1 The peacock butterfly. Photo: Niklas Janz.

There is a large patch of stinging nettles, *Urtica dioica*, that will dominate the southeast corner of the garden later in the summer, but which at this time of the year is only visible as some small green tufts protruding from the soil. The nettles don't really belong in a nice garden, but they are hard to get rid of, and Mary doesn't really mind. Now she watches the solitary insect as it searches the garden, waiting for it to find its way to the nettle stand. There! Now the butterfly lands much more frequently, evidently looking for the perfect place for its offspring. Eventually it is down for quite a while, and Mary gets up from her chair, down from the porch, and walks out on the lawn, slowly approaching the insect in order to keep track of the exact spot. When the butterfly finally alights, and then quickly flies high up above the bushes and hedges and leaves the garden, Mary knows precisely where to look. She walks up to the spot and brushes some wilted grass to the side, exposing a tender shoot of nettle. And sure enough, as she flips the nettle leaves over one by one she soon finds a batch of what looks like at least a hundred small green eggs on the underside of one of the leaves. In a couple of weeks they will hatch, and a group of thorny black larvae will ravage the nettles that by then will have reached a much more considerable size.

Mary is pleased. Out of all the plant species in the garden, the peacock once again chose to lay her eggs on stinging nettle, just like every time in previous years – or indeed at all other places where she has been able to study the egg-laying behaviour of this particular species. Mary feels confident that one aspect of the world is known and predictable:

Butterflies of the species *Aglais io* specialise on the plant species *Urtica dioica* as food for its larvae.

2.2 Early Empiricism: Francis Bacon

As we will see throughout this book, it is not a simple task to define precisely what science is, and how it differs from other human activities. To begin with, we can safely say that *science is an activity with the goal to produce knowledge*. As we mentioned in Chapter 1, there are other ways to obtain knowledge, but among them, science has a special standing in the modern industrialised world. Consequently, there have been many attempts at formulating a philosophical theory of what science is and how science should be done.

Returning to Mary's story, first, how can Mary know that her sightings of this particular species laying eggs on nettles are representative of what this species always does? After all, it's *possible* that they do something else whenever she isn't looking. Perhaps they prefer to lay eggs on crocus leaves in the morning before Mary is up and out, and only later in the day switch to nettles? True, she has only seen larvae feeding on nettles, not on any other plant species, but perhaps the larvae move from crocus to nettle when they are still small enough to not be easily seen, making a bold and dangerous trip across the lawn? Yes, this is possible, but it is a more far-fetched and complex theory than the idea that the species is a nettle specialist, and one of the oldest scientific principles is that we should stick with a simple theory unless there is good evidence that it might be wrong. In other words, if there are competing hypotheses equally supported by data, we should select the one with the fewest assumptions. This has been called the law of parsimony or Occam's razor (from Franciscan monk William of Ockham, 1287–1347). It remains a sound principle to this day, not because nature is always simple but because the alternative leads nowhere. If you don't care about this aspect, there are always innumerable possible hypotheses that fit the data. We would never agree on anything, and there can be no scientific progress.

Mary has additional reasons to accept the generality of her findings. The fact that she has seen peacock females laying eggs on nettles in several different years, but never seen them lay on another plant species, is an indication that it is a general rule. But maybe it's just in her garden that they prefer this species as a larval host? No, she has in fact seen the same behaviour elsewhere on her travels. Evidently, the observation that this species is a nettle specialist has been repeated over time and in different situations, and is (at least to Mary's knowledge) not contradicted by opposing evidence. So, it can at least provisionally be formulated as a rule – as knowledge about the world.

Mary's intellectual process can serve to illustrate the first attempt at a philosophy of science. It goes by many names in its various versions, notably *empiricism* or *inductivism*, and, later on, *positivism*. The general idea is this: by carefully and objectively observing the world, we will obtain verifiable facts, and from these facts, we will see patterns emerge. If we always see peacock butterflies laying their eggs on stinging nettles (and not on any other plants), we can make a generalisation: this species of butterfly always lays its eggs on this plant species, even when we are not there to observe it. Such a generalisation is an example of the process of *induction*, when we extrapolate from a limited number of observations to produce a general rule. Once we have such rules, we can use them to make predictions in a process of *deduction*: we can predict that the next time Mary or somebody else sees a female peacock lay its eggs, even in a different year and place, it will certainly do so on a stinging nettle. Why? Because this is what this species does. It's a rule.

Such generalisations are second nature to us as humans, or even as animals. We simply can't help trying to see patterns in the world around us: if crocus is one of the first flowers that we see in the spring, a couple of years in a row, we expect this to happen next year as well. And buds will break, and spring butterflies will show up. It would indeed be impossible for us to function without such generalisations, because the world would appear completely unpredictable. How would you know that the bus stop you go to every day is likely to be in the same place tomorrow? Why not set out in a random direction when you leave home, since the bus stop could be anywhere?

Animals do it, too. As everybody who has ever had an aquarium can attest, fish quickly learn to associate the sound of a food jar banging against the side of the aquarium with being fed and will swarm to the surface when they hear it – in other words, they are generalising from earlier occasions to the next. The fish now 'know' something about the world: that particular sound means food. When it comes to animals, we don't necessarily believe that any advanced thought processes are involved, but as humans, we are capable of not only generalising inductively but also articulating theories about explanations – and sharing these ideas with others, as a first step to science.

It is only natural, then, that inductive reasoning based on empirically obtained facts was an important part of the first more modern attempt at formulating a theory of science, which can be attributed to Francis Bacon (some of the 'ancient Greeks' had similar ideas, but we will not explore them here, as we are primarily concerned with science as it is done today). Bacon (1561–1626) was an English statesman and philosopher active during the times when modern science was born – the so-called scientific revolution – and he is sometimes referred to as the 'father of scientific method'. He stressed strongly the importance of careful empirical observations and experiments without prejudice as the only way to obtain truly new knowledge, rather than obtaining knowledge from ancient authorities or metaphysical speculation.[1]

---

[1] Bacon, F. (1620) *Novum Organum, Aphorisms-Book I (I and XXXI)*.

Man, as the minister and interpreter of nature, does and understands as much as his observations on the order of nature, either with regard to things or the mind, permit him, and neither knows nor is capable of more.

It is in vain to expect any great progress in the sciences by the superinducing or ingrafting new matters upon old. An instauration must be made from the very foundations, if we do not wish to revolve forever in a circle, making only some slight and contemptible progress.

Moreover, Bacon advocated the equally sound scientific principles that you should not jump to conclusions and not generalise beyond what is reliably demonstrated by the facts that have been obtained from observations, and you must be prepared to abandon your provisional generalisations when confronted by opposing evidence. He cautions that this goes against human nature, so there is a need for 'severe regulations' in the scientific process.[2]

2.2.1 The Baconian Method

Bacon outlined a method for how to impose these regulations – how to proceed from particulars (observations and simple experiments) to what he called lesser axioms (generalisations covering the particulars and not much else), to 'intermediate axioms' (more general rules) and finally to 'general axioms' (laws of nature). This has to be done in careful steps, checking, for instance, that proposed intermediate axioms are indeed successful in 'pointing out new particulars', that is, in making predictions beyond what has already been observed. Once we have a reliable axiom, we can use it to make predictions, or in Bacon's words: 'the deducing or deriving of new experiments from axioms'.

Bacon strived to start the process of obtaining knowledge over from the very beginning, rather than trusting in ancient philosophers like Aristotle. To this end, he demonstrated his method with an attempt at a 'natural history' of the world. He would take a phenomenon such as 'heat' and tabulate all instances where heat is found (e.g. rays of the sun), along with a table of comparable instances where it is not found (e.g. rays of the moon). The strength, or degree, of correlation with the phenomenon is also important: rays of the sun are more associated with heat in summer than in winter, more at noon than at other times of the day, and more near the equator compared to near the poles. If you approach a hot object, the sense of heat is not constant but increases proportionately. Such observations were painstakingly tabulated by Bacon, along with suggested additional observations and experiments needed to clarify if the axioms really hold true. These tables, Bacon suggested, can be used to reveal the nature of a phenomenon such as heat through a process where we first reject all hypotheses that are contradicted ('every contradictory instance

---

[2] *The human understanding is most excited by that which strikes and enters the mind at once and suddenly, and by which the imagination is immediately filled and inflated. It then begins almost imperceptibly to conceive and suppose that everything is similar to the few objects which have taken possession of the mind, while it is very slow and unfit for the transition to the remote and heterogeneous instances by which axioms are tried as by fire, unless the office be imposed upon it by severe regulations and a powerful authority (Novum Organum I: XLVII).*

destroys a hypothesis as to the form') and finally look for conspicuous positive evidence revealing the nature (or 'form') of the phenomenon.

Heat is not purely a material or terrestrial phenomenon, since the rays of the sun produce heat. Neither can it be purely of celestial origin, since fires and even subterranean fires are hot. And so on. Bacon concluded instead that *motion* is the common denominator of all instances where heat is found: flames move; boiling liquids move; heat can destroy substances by causing violent internal motion; heat makes substances expand (i.e. their parts move outwards from the centre, whereas cold makes them contract). Thermometers are particularly conspicuous clues to the nature of heat because they show the precise degree of heat as a consequence of such expansion and contraction. Bacon came up with a definition of heat that is at least superficially reminiscent of a modern explanation of heat as being the kinetic energy of atoms or molecules: 'Heat is an expansive motion restrained, and striving to exert itself in the smallest particles.'

2.2.2 Did Scientists Historically Follow the Baconian Method?

The precise 'Baconian method' of tabulating facts (much simplified in the description in Section 2.2.1) was probably not followed by many, but his general ideas were quite influential among scientists for centuries, and to some extent still are. When Isaac Newton proclaimed *Hypotheses non fingo!* (approx. 'I contrive no hypotheses') in the 1713 edition of his *Principia*, going on to state that 'hypotheses […] have no place in experimental philosophy' (although he clearly articulated hypotheses and tested them all the time in his own work), he evidently meant this in the Baconian tradition: hypotheses (Newton preferred the term 'propositions') must be close to the facts and not arbitrary or speculative, or otherwise they will only bias the mind of the scientist.

Even later, Charles Darwin, in his autobiography, felt the need to state that he had 'worked on true Baconian principles and without any theory collected facts on a wholesale scale'[2]. This may possibly have been true of his early career, when he travelled on the *Beagle* and collected specimens and observations that he later made use of to articulate his theory of evolution by natural selection. However, as we shall see in Chapter 3, it is more probable that Darwin used an approach closer to modern hypothesis testing from the beginning. In his autobiography, he also wrote that 'I cannot resist forming [a hypothesis] on every subject'. This indicates that Darwin's actual scientific process was far from collecting facts 'without any theory'. It's not hard to imagine that Darwin was apprehensive that his ideas about evolution might be dismissed as speculation, and that he wanted to emphasise his objectivity and careful scientific process in this way.

2.2.3 Bacon's View of Progress

One could argue that progress is a central goal of science; otherwise, it would make little sense. It is therefore worth considering how such progress would look, according to the different philosophies we meet. What would scientific progress be for an empiricist? Bacon's view of progress could perhaps be seen as a gradual increase in generality of the axioms that the method uncovers. If the method works, each axiom that is arrived at should be considered a truth and should not be expected to later turn out to be false. So, unless methodological errors have been made, progress should not involve replacing one axiom for another but rather inducing more and more general 'truths'.[3]

2.3 Later Empiricism: The Positivists

In the times since Bacon, empiricism has taken on different shapes and forms, and one of the most influential of these developments is what has been called positivism. In a general sense, positivism is simply the notion that useful scientific knowledge must be derived via reason and logic from empirical observations, much as what Bacon proposed, with 'positivity' referring to the degree of exactness with which phenomena can be measured and determined. The history of positivism as a philosophy of science is complex, and we will not go into all its twists and turns here.

Positivism is considered to have been first articulated by the philosopher of science Auguste Comte (1798–1857), who was one of several thinkers in the early nineteenth century attempting to apply the successful scientific method of the natural sciences to other fields of research. Comte wrote a series of volumes dealing first with the natural sciences, and then with the emerging field of sociology. He acknowledged mathematical demonstrations as how you can gauge the exactness (positivity) of a science, ranking different fields in order of decreasing positivity: astronomy, physics, chemistry, biology and sociology. Comte's influence extended to politics because of his theory of social evolution. This account proposes that society in its search for truth about the world must go through a progression of three stages, from the 'theological'

---

[3] From *Novum organum 1: XIX*: '[Inductivism] constructs its axioms from the senses and particulars, by ascending continually and gradually, till it finally arrives at the most general axioms'.

(accepting doctrines of the church), over the 'metaphysical' (when logical rationalism and universal rights of humanity prevail) to the 'positive', scientific, stage (a utopian stage when rights of the individual and free will are most important). These ideas, together with Darwin's, influenced the rise of several secular humanist organisations. Other brands of 'positivism' soon developed; notably, Émile Durkheim (1858–1917) outlined how the principles of natural science (objectivity, rationalism and causality) could be applied to the social sciences, and other variants of positivism influenced the study of history, psychology and economics, with the general aim of explaining and predicting patterns through generalisations from positively observed facts.

Logical positivism was a very influential brand of empiricism with its main roots in Vienna between the World Wars, when groups of philosophers, scientists, and mathematicians met and formed a discussion circle. The central thesis of this philosophy (based on Wittgenstein's early philosophy of language) was the criterion of 'verifiability', a theory of knowledge asserting that in order to be meaningful and scientifically relevant a statement must be verifiable through direct observation (sensory experience) or logical proof (as in mathematics). Unobservable structures and forces thus have no place in science. But what about, for instance, physics and its (very successful) use of terms referring to phenomena that cannot be directly observed, such as 'atoms', 'charges' or 'nuclear forces'? One brand of logical positivists tried to deal with such unobservables by viewing them as purely logical constructs aiding in finding observable truths about the world, rather than literally referring to (unobservable) entities. If matter behaves 'as if' it is made up of atoms, this can be a useful idea, but only to the extent that it leads to predictions that can be verified by observation and measurements. It does not necessarily mean that atoms are real. After all, 'atom' is an old concept, even dating back to Ancient Greece, but the meaning of the word has changed considerably with the emergence of new theories (see also Section 9.1.4 on models in science), and conceivably it may change again following future discoveries. In what sense can we then say that atoms are real? Logical positivism is thus a distinctly 'antirealist' view of science. A realist would instead argue that when the meaning of 'atom' changes as new knowledge emerges, it does so precisely because it gives an increasingly better match with the real world.

2.4 The Limitations of Empiricism

In its simplest form, empiricism amounts to making a number of straightforward observations of natural phenomena using our senses, and then through a process of induction make general statements about the nature of the world based on observations. Once in place, such general statements (or 'laws') can be used to deduce predictions of, and explanations for, natural phenomena. There are a number of problems with this proposed scientific process that have been pointed out by later philosophers of science.

2.4.1 Logical Problems

First, there is the logical problem that a finite number of observations can never be used to form a universal statement, one that is certain to hold for all cases. We may heat up one bar of metal and note that it expands, and then make the same observation using another bar, but this does not necessarily mean that all metals will always expand when heated. We can try to reduce this problem by making many observations of many different metals and varying the circumstances, but there could still be exceptions out there that we haven't yet found. We can never prove the truth of a universal statement, at least not outside of pure mathematics.

David Hume (1711–1776), another early empiricist, famously first formulated the 'problem of induction', namely that there is no rational reason for believing that the future will always resemble the past, or in other words, that we can make predictions or infer causality based on past observations. The only way to justify that the principle of induction works is because it has worked in the past; that is, to justify induction, we have to appeal to induction. This problem of circularity did not stop Hume from acknowledging that induction is important as a principle and something all of us make use of, but he was of the opinion that the reasons for using the principle are not rational but rather stem from experience, custom and mental habit.

A quite different logical problem is the so-called raven paradox. Suppose that you, through inductive reasoning, have reached the hypothesis that all ravens are black, and that you wish to further strengthen it. What kinds of observations should you set out to make? It seems reasonable to think that you should go out looking for black ravens, but the raven paradox questions the logic behind this intuition in a way that indeed feels paradoxical.

The statement *all ravens are black* is logically equivalent to its contrapositive: *all non-black things are non-ravens*. Hence, the hypothesis would be supported by an observation of a black raven, but simple logic dictates that it would also be supported by observing any non-black object that is not a raven, such as a green umbrella. This does seem counterintuitive, but there is nothing in the inductive machinery that would allow an empiricist like Bacon to argue why observing a black raven supports the hypothesis any more than observing a green umbrella, because the two statements *all ravens are black* and *all non-black things are non-ravens* contain the exact same information.

2 Observation-Driven Science

Logical positivism, the later development of empiricism that we mentioned in Section 2.3, fares no better in this regard. The paradox exposes a tension between the logical requirements of logical positivism and our intuition about the types of evidence that are actually relevant and meaningful. In our example, formal logic requires that the two observations must provide equal confirmation, but our intuition suggests that observing a black raven supports the hypothesis, but observing a green umbrella does not. Interestingly, the paradox was first formulated by Carl Gustav Hempel, one of the main proponents of logical positivism [3]. He recognised that the paradox was a real problem for this standpoint, and while it did not cause him to immediately abandon his position, it may have contributed to him eventually moving away from it later in life. The paradox teaches us that a good scientific approach also needs to account for what constitutes a relevant observation.

2.4.2 Observations Are Fallible

Second, as we saw in Chapter 1, observations are seldom or never straightforward, objective and theory-free. Our senses cannot really be trusted to always perceive the objective truth, as can be demonstrated, for instance, by any number of optical illusions (Figure 2.2).

Figure 2.2 Example of an optical illusion – the vertical lines are actually straight. Observations are not always straightforward and dependable.

If we instead use some sort of instrument to make the observations, they are no longer direct but dependent on theoretical assumptions regarding the workings of the instrument. Also, using an instrument does not necessarily make the observation objective. There will always be a tendency to see what we expect to see, introducing subjectivity. Furthermore, what we observe will be dependent on our knowledge and training. Any student who have studied preparations of various tissues in a microscope can attest that it can take quite some time to learn what they actually show. Once learnt, the patterns and structures that at first may seem alien and strange suddenly make sense and start to form cell walls, vacuoles, chloroplasts, etc. At that point, detecting these features seems easy and natural, but it is a professional skill that typically requires hard work to learn. Thus, 'observing' is not something you can passively leave to your senses; it is a complex process of interpretation. Similarly, a tourist visiting the Grand Canyon will see a spectacular sight: a wide and deep gorge flanked by beautiful steep sides of rock, striped in different colours (Figure 2.3). Far below at the bottom of the gorge, there runs a meandering river. With a little education in the geosciences from school, you would recognise the stripes as representing layers of rock laid down as sediments during different geological

Figure 2.3 An aerial view over the gorge at Grand Canyon National Park, showing the striated walls. Where most tourists see only a spectacular sight, an observer with an education in the geosciences would be able to deduce some of the geological history of the area. Photo: Carol M. Highsmith. Photographs in the Carol M. Highsmith Archive, Library of Congress, Prints and Photographs Division.

periods, with the youngest nearest the surface. You would perhaps also realise that the canyon was likely carved out by the river (or by some earlier stream of water). A trained geoscientist exploring the canyon more carefully by going down into it and doing some digging will notice much more, including what is known as the Great Unconformity: rocks below the bottom of the canyon are tilted (non-horizontal layers) and very much older than the striped sides of the canyon. There is thus a huge gap in the geological layers, hypothesised to be due to massive erosion connected to ancient glaciation events. The tilted rocks are evidence of past dramatic geological activity in the area. Adding even more theory, trace fossils and radioisotope dating can be used to date the rock layers.

It is instructive to consider in depth any example of actual observations made in the context of a scientific investigation or a student project and outline the various ways in which such observations are not really theory-free. When performing this exercise in class, we have found that student projects in biology fall somewhere on a spectrum from relatively straightforward observations of animals or other living organisms (for instance, an animal behaviour study or an investigation of plant community composition) to highly theoretical studies in fields such as protein crystallography or genomics. In the latter case, the theory-dependence of observations is very evident. The observations as such will be the output from some sort of instrument and can only be interpreted using complex theory. Often, the aid of computers is absolutely necessary to even visualise results, let alone interpret and analyse them. A similar situation applies in scientific fields such as particle physics or gamma-ray astronomy, for example.

However, theory-dependence is ever-present also in the former type of studies. For one thing, observations of organisms are dependent on theoretical constructs such as species, populations, communities, or the classification of the behaviours studied. The 'proteins', 'genes', or 'particles' studied in the more obviously theoretical fields are also theoretical constructs, of course, and such words can in fact fruitfully be seen as shorthand for entire theories, changing over time as the field of study develops. When the word 'gene' was first coined in 1909 by the Danish botanist Wilhelm Johanssen, it referred to the Mendelian unit of inheritance [4]. This was an abstract concept, since the mechanisms for inheritance were not yet known. Today it can still be used in this way in, for instance, quantitative genetics, but more commonly it refers to the physical gene, a string of DNA. Similarly, the concept of what is a 'species' has evolved over the years and is still a matter of much theorising. Thus, even when we are 'simply' investigating the species composition in a community, we are in fact dependent on theory regarding what a species is, and which organisms should be considered as belonging to which species. Another theoretical aspect concerns why the investigation was performed in the first place. It has been our experience that in biological fields such as ethology or ecology, the observations themselves can be relatively straightforward, but they are typically made in the context of testing complex theory. Say that we are simply counting spiders in a pitfall trap for an ecological study. Ecology is a theory-heavy field and there is typically a large amount of theory required to motivate the observation in the first place – Why is it something worthwhile to do, and why in this particular way? Why count spiders and not something else? Why use this particular type of trap? Why in this spot and not somewhere else? The opposite is often true in, for instance, molecular biology, where the observations are highly dependent on theory but are made in order to answer relatively straightforward questions such as what is the function of a particular gene product.

Why does it matter if observations are not theory-free? It matters because it means that they are always fallible. Scientific observations, even published ones, can later be found to be in error – for any number of reasons. For instance, any measurements taken can be faulty, especially when they rely on theoretically complex machinery for detection or analysis, but also when measuring tools are simply not up to the task. This must be the reason why Bacon wrote that iron does not expand when heated and even used this 'fact' as an example of an observation that can be used to reject a general law, in this case the 'absolute expansive mode' of heat.[4] More commonly, however, one or more of the theoretical assumptions used to interpret the observation or design the experiment was in some respect faulty.

Consider even a simple observation like the ones made by Mary in her garden. Studying egg-laying on a particular plant is interesting if we assume that this will be the plant used by larvae as food. This is, in many cases, a reasonable theory, but there are in fact examples of butterflies that don't lay their eggs on the host plants but instead on nearby tree trunks. This means that it could be the case that peacock larvae crawl away from the nettles after hatching, to find their real host plants (they don't, but Mary couldn't know this for sure without investigation). What if this species does use nettles, but also lay their eggs in other places that are harder to observe, such as high up in trees? What if what Mary believes to be the peacock butterfly on closer investigation would turn out to be not a single species, but several very similar ones behaving somewhat differently? What if the plant known as stinging nettle instead turns out to be more than one species, and not all of them are accepted by peacock females?

---

[4] *Novum Organum* (Aphorisms Book 2: XVIII); 'Exclusive Table, or of the Rejection of Natures from the Form of Heat'.

How do we know which past observations can be ascribed to which species? These are just a few examples of how scientific observations are always dependent on theory and could turn out to be entirely or partially wrong. The risk of this happening is naturally even greater in more theoretically complex fields of research, where it is not uncommon that published scientific papers later have to be corrected or even completely retracted when the authors realise the shortcomings of the results.

2.4.3 Observing without Theory Is Ineffective

The proposition of simplistic empiricism that initial observations should be theory-free in order to be objective also leads directly to another limitation. Without theory, how can we know which observations to make, and how to make them? It is simply not an efficient way to conduct science, and in Chapters 3–4 we will see how having a hypothesis to test before conducting observations ensures that the relevant observations are actually made.

A related limitation of such an approach is also most evident when compared to hypothesis testing. Heating up a number of different bars of metal and noting that they all expand could lead to the general statement that 'metals expand when heated', but *why* do they expand? In science, we strive to understand the causal mechanisms behind natural phenomena, and there is not much in empiricism to guide us in this direction. As noted in Section 2.2, there is a deductive part of empiricism where explanations are derived from general statements, but without adding much other knowledge (as in Bacon's painstaking tabulation of phenomena related to heat, see Section 2.2.1), this will only be explanations of the form 'the bar expands when heated because it is a metal, and we know that metals expand when heated'. This is hardly satisfactory as a scientific causal explanation.

2.5 The Place of Empiricism in Science

Bacon's ideas should be understood as a necessary and important reaction against the situation in his time, when philosophers could speculate freely from little data and Aristotle's works were still upheld as the truth about nature. It is also an intuitively appealing view of the scientific method, and indeed, it seems to be a common perception among non-scientists that this is precisely how science works. A police detective in a TV show may claim to use scientific principles by insisting on objectively collecting lots of evidence before forming any theory on who might be a murder suspect. In reality, police will certainly first suspect the male partner if a woman is found dead in her home and check his alibi before that of any random guy on the street. Similarly, as we saw in Section 2.4, scientists always have theories governing observations – what to observe and how to observe it. In Chapter 3, we will see how this often amounts to making observations in order to test specific hypotheses, rather than collecting just any 'facts' before theorising.

However, there is a sense in which empiricism has a central place also in modern science, and many scientists would describe themselves as empiricists first and foremost. For some, this just means that they are not theoreticians; they make actual observations and experiments. Others would probably take it further; they use these observations and experiments to collect scientific knowledge, and they do not always have a hypothesis to test in mind. In applied sciences, this is a common situation, because the researcher may be looking for useful data rather than trying to change theory in the scientific field. This could involve just measuring something more precisely than has been done previously in order to improve on the applications of science. Another, currently expanding, scientific area where the empiricist approach seems particularly obvious is what is referred to as 'Big Data' or 'data-driven science', that is, the development of new methods (such as Artificial Intelligence, AI) to analyse huge datasets and find patterns that are not immediately obvious or even tractable with standard methods. Because of the nature of such problems, it is often seen as an advantage that the analysis is 'unbiased', that is, not constrained by prior theory (echoing Bacon and other early empiricists). Such examples suggest that the limitations outlined in Section 2.4 may not always be decisive. Let us look at them again.

The logical problem that universal statements cannot be derived from a finite number of observations is, in a sense, crucial only for philosophers. In science, it is important to attempt to make generalisations, as we cannot study every single case of anything. There may well be exceptions to these generalisations, but they can still be very useful summaries of current knowledge if they hold in most cases. An exception, rather than invalidating the generalisation, can also serve as a starting point for better understanding of the phenomenon – why does the generalisation usually apply, but not in this particular case?

The other limitations do amount to strong objections against at least the most simplistic brands of empiricism. Observations are not theory-free, and insisting on trying to make them without theory can be very inefficient and will not easily lead to causal explanations. However, this does not exclude a role of empiricism-like philosophies as part of science. Useful observations can sometimes be made without a well-formulated hypothesis, and they may aid in the formulation of such hypotheses. Moreover, the inductive process is indeed how we make generalisations, as in the example of Mary's butterflies.

There have, in fact, been more recent developments in the philosophy of science grounded in empiricism. 'The new experimentalism' promoted by Deborah Mayo (among others) claims that there are experimental results (and associated knowledge) that stand even if the theories that explain them change [5]. A simple example could be an electric circuit with a battery and a lamp, where the lamp lights up as the circuit is completed. This outcome remains true even if our theoretical understanding changes regarding exactly what electricity is, or how the battery functions, and any new theories concerning electricity must be able to explain why the lamp lights up in a closed circuit but not when it is broken. Such results approach the 'facts' accumulated in science according to the earliest philosophies of science, the empirical approaches. For less simplistic examples, however, Mayo acknowledges that experimental results are always theory-dependent to some extent, and she outlines how experiments should therefore ideally be carefully designed and results strictly analysed so that the outcome can have only one interpretation, in this way adding knowledge.

3
Hypothesis-Driven Science: Falsificationism

Science is based on critical testing of hypotheses about how the world is.

3.1 Mary's Butterflies

Let's see what Mary has been up to since we last met her:

> In the days after Mary's encounter with the peacock female, spring took a step back and the weather was too cold and damp for butterflies. However, as soon as the sun started to warm up her garden again she was out there, eager to see if any of the other early flying butterflies would show up.
>
> It was not long before she saw a flicker of orange dashing into her garden from over the birches and willows, starting to patrol the area in a peculiar skittish flight from plant to plant. It was more lightly coloured than the peacock, and this, together with the flight pattern, suggested to Mary that this was another species. This was the beautiful comma butterfly – *Polygonia c-album* (Figure 3.1).
>
> Mary sat down to watch, and could soon take note of the butterfly exploring the very same nettle patch where the peacock had laid her batch of eggs. Mary knew that one difference between the species is that females of the comma don't lay eggs in large batches, but rather only one or a couple of eggs at a time. Thus, she was not surprised when she saw the butterfly landing for just a few seconds on a nettle leaf and then taking off again. Quite possibly it had already had time to lay an egg, and was now searching for the next site for her offspring. Mary was feeling a bit drowsy in the sun and did not even bother to get up to check. Almost every spring she had sooner or later found the conspicuously coloured comma larvae on her nettles, so she was already confident that this close relative of the peacock was another specialist on using *Urtica dioica* plants as food for its larvae.
>
> But what was this? The comma did not move on to another nettle leaf or plant, but instead made a dash for the black currant bushes at the edge of the garden. She saw it land on a currant leaf near the tip of a branch. Maybe it was just going to bask there for a while, warming up before continuing her search for suitable host plants?

3 Hypothesis-Driven Science: Falsificationism

Figure 3.1 The comma butterfly. Photo: Niklas Janz.

If so, it would spread its wings to maximise the surface exposed to the rays of the sun. But this was not what happened. Rather the butterfly immediately folded its wings and started to crawl around in the typical manner of a female checking out a potential host by its texture, odour and taste.

Mary got up and walked quickly towards the bush, excited to see if indeed an egg would be laid on this unexpected plant. However, in her eagerness she accidentally got between the sun and the butterfly, casting a shadow that immediately prompted the butterfly to take off and propel itself straight up and out of her garden. Disappointed, Mary decided that she could not wait for the next time she happened to observe a comma female exploring her currant bushes, as this might be years from now. In all likelihood she had just misinterpreted her observation, but was there a way to check?

After some thought, she got out her butterfly net, and waited. And waited. It was not until two days later that she finally saw a comma in her garden again, behaving suspiciously like a female among the plants in the nettle patch. Mary approached it cautiously, and when the butterfly was busy investigating a nettle leaf, she saw her chance. With a quick swing of the net and a twist of her hand she had the insect safely collected.

Up on her porch she had already prepared a simple experiment. In a big jar she had placed a twig from her black currant bush and a small nettle plant, matched more or less to size. Some damp paper at the bottom of the jar provided some moisture that hopefully would keep the plants fresh for a while, as well as something for the butterfly to drink from. She carefully fished out the comma butterfly from her net with her right hand, placed the insect inside the jar, and before releasing it used her left hand to seal the jar with a piece of gauze. A string around the tip of the jar to fasten the cloth finished off the set-up. Mary placed the jar on a small table, set in a sunny part of the porch beside her chair.

Now Mary could only wait again. But it was not long before the butterfly had warmed to the situation and started to crawl around among the plants. Mary wondered why she had never thought of this experiment before – it was very interesting to be able to see the egg-laying behaviour up close, while sitting comfortably in her chair. She could see the female curling her abdomen to reach down under a nettle leaf, depositing a single green barrel-shaped egg. And soon again, on another leaf. And again. But then it happened! Using exactly the same behaviour, the female laid an egg under a currant leaf.

Apparently, Mary had been wrong in her earlier hypothesis. *Polygonia c-album* is not a strict specialist on *Urtica dioica*, but instead at least occasionally accepts other host plants, at least black currant, *Ribes nigrum*. Mary was forced to adjust her view of the world, and could not wait to write to her fellow butterfly enthusiasts to share her observation. Maybe someone else had observed the same thing, or even egg-laying on yet other plants?

3.2 Karl Popper and Falsificationism

Mary realised that pure observation would not be an efficient way to answer her question. She was not the first to have this realisation. Other more prominent scientific figures struggled with the inefficiency of empiricism as well. In Chapter 2, we hinted at Darwin's struggle with the Baconian method. Let us have another look at what this entailed. On the first page of his *Origin of Species* [6], Darwin wrote:

> On my return home, it occurred to me, in 1837, that something might perhaps be made out on this question by patiently accumulating and reflecting on all sorts of facts which could possibly have any bearing on it. After five years' work I allowed myself to speculate on the subject.

He clearly wanted to show the reader that he followed the proper scientific method by refraining from formulating hypotheses until after he had accumulated enough observations. But did he really? In his more private conversations, he gave air to a rather different view:

> About 30 years ago there was much talk that Geologists ought only to observe & not theorise; & I well remember some one saying, that at this rate a man might as well go into a gravel-pit & count the pebbles & describe their colours. How odd it is that every one should not see that all observation must be for or against some view, if it is to be of any service.[1]

It appears that Darwin's explicit nod to the Baconian scientific method was largely lip service, something he needed to write to be taken seriously in the prevailing scientific climate. Indeed, in a letter to a young scientist, Darwin wrote:

> Let theory guide your observations, but till your reputation is well established, be sparing of publishing theory. It makes persons doubt your observations[2]

---

[1] Letter to Henry Fawcett (18 September 1861).   [2] Letter to John Scott (6 June 1863).

It would seem then that Darwin was as troubled as Mary with the Baconian method, but he thought it best not to speak about it. There is probably a lesson to be learnt here about the perils of scientific dogmatism, but it is also interesting to note that working scientists were not completely constrained by the philosophers' shackles. Nevertheless, the insistence of gathering substantial amounts of 'objective facts' before hypothesising about a problem prevailed as an ideal in natural sciences long after Darwin's book on the origin of species.

3.2.1 The Logic of Scientific Discovery

The person who finally convincingly countered this view of science was the Austrian philosopher Karl Popper. During Popper's student days, he was exposed to the influential 'Vienna circle' of logical positivists that we met in Chapter 2, and he reacted against the same issues that we have seen Mary and Darwin struggle with. Popper took issue with the positivists and how they claimed that theory should be built from careful unprejudiced observation. He pointed out that this is logically flawed, since a hypothesis can never be logically deduced from observations, however many they may be. Induction, inferring general truth from specific observations, is not a logical process. 'All swans are white' does not *logically* follow from observing any number of white swans, as we can never be certain that a black swan may not turn up the next time we look. As discussed in Chapter 2, empiricists were well aware of this but did not consider it a fatal flaw. After all, if a theory is thoroughly induced from a large number of observations, it is very likely that it will hold true in the future as well. Popper was not as forgiving, and his alternative approach rested on an important asymmetry in the relationship between universal and singular statements: while universal statements cannot logically be deduced from singular statements, they can be contradicted by them. The hypothesis 'all swans are white' is logically contradicted by a single observation of a black swan. Popper called such a contradiction a *falsification* [7, 8].

3.2.2 A Shift in Perspective

The advantage of falsification over induction isn't just about logic. As we have seen, both Mary and Darwin noted that the inductive process is inefficient and wasteful as a means of gaining scientific knowledge. Inductive inference is based on generalisation. Adding further observations will always strengthen this inference, but to a diminishing degree. When Mary observed a comma butterfly lay another egg on a nettle, it gave support to the hypothesis that it is a nettle specialist, and any subsequent similar observation would further strengthen this hypothesis, but less and less for each such observation. In the end, she didn't even bother getting up to verify the observation. However, the potential observation of the comma butterfly laying eggs on another plant species would immediately refute the hypothesis that it is a nettle specialist, and quite understandably this got her up on her feet! A scientist who has tried to publish a result that is 'already demonstrated' also knows that subsequent documentation of the same observation is typically not considered very interesting by their peers. If you, after observing a great number of swans, formulate the theory 'all swans are white', at some point, it would feel increasingly meaningless to keep searching for white swans to further strengthen this theory. However, if a single observation of a black swan would be enough to demonstrate that the theory is false, wouldn't it make more sense for the scientist to actively set out to look for swans that are not white? Consequently, good scientists should not strive to verify their hypotheses; they should strive to falsify them.

This is a dramatic shift in perspective. According to Popper, the goal of science is not to formulate hypotheses (however well founded), but to *test* them. A hypothesis can only be empirically tested after it has been advanced, and quite contrary to what the empiricists taught, theory must then precede observation. An observation is only meaningful if it can be used in testing a hypothesis. Falsificationists thus fully embrace the theory-dependence of observations that we found to be problematic for the observation-driven approach to science. Theory-dependent observations are only problematic if you claim that observations must not presuppose theory, and in falsificationism, observations are very much guided by theory. Theory is what makes an observation *interesting*.

If a naive observer saw a comma butterfly land right in front of them, they may make the observation that the butterfly has four legs. If the observer was truly naive, this would be a rather uninteresting observation, on par with observing that the butterfly has two antennae or two eyes. But if the observer had some basic biology training, they would realise that this is in fact a very odd and unexpected observation, since butterflies are insects and all insects should have six legs. Hence, the observation that the comma butterfly has four legs is interesting *in light of the theory* that states that it should have six. This theory might even prompt this observer to further investigate the butterfly, which would reveal that the butterfly does indeed have six legs, but the anterior pair is repurposed, and held so tightly against the body that they are scarcely visible (see Figure 3.2). Most observations are pointless in themselves. If you don't know what you are looking for you might indeed as well 'go into a gravel pit

3 Hypothesis-Driven Science: Falsificationism 33

Figure 3.2 A comma butterfly, the regular legs, and the barely visible two front legs that are held tight under the front of the body and not used for walking (arrow). At first glance, this species seems to have only four legs, which from current theory is unexpected for an insect. Photo: Niklas Janz.

and count the pebbles and describe the colours', as Darwin pointed out in the beginning of this chapter.

An important tenet of falsification is that for a hypothesis to be falsifiable, it must be possible to formulate potential outcomes that would *not* be in agreement with the hypothesis. Popper would argue that this is the purpose of hypothesis testing; by stating what would be expected if the hypothesis is true, we must also accept that there are other possible observations that could be made that would not be in agreement with the hypothesis and thus falsify it. This is what it means that a hypothesis is falsifiable, and a good scientist must be prepared to abandon the hypothesis if such an observation is indeed made. If so, the hypothesis should be considered falsified and placed on the scientific scrap heap, and the scientist should move on to other, perhaps more successful, hypotheses. For Popper, the principle of falsifiability was so important that he claimed that it was the very criterion by which to define science. He considered this to be a major problem with the methods that relied on induction, because we use induction in all kinds of everyday situations that we don't consider scientific. Hence, induction does not provide a clear way of distinguishing science from non-science. Popper considered this 'problem of demarcation' (between science and non-science) to be essential to any theory of knowledge, and he claimed to have solved it. What clearly distinguishes science from other activities in society is that science produces falsifiable hypotheses and abandons them when they are falsified. To be a Popperian scientist then involves three characteristics: an ability to formulate falsifiable hypotheses, a willingness to ruthlessly test them and a willingness to abandon hypotheses that are falsified.

3.2.3 Falsifiability

How do you distinguish a falsifiable hypothesis? Here are some examples of hypotheses that are falsifiable:

> All swans are white
> Female comma butterflies always lay their eggs on leaves of the stinging nettle
> The sun always rises in the west
> The positive and negative ends of two magnetic bars will attract each other

The first hypothesis is falsified by the observation of the Australian bird *Cygnus atratus*. We saw earlier in this chapter how Mary falsified the second, and the third is easily falsified if you wake early enough on any given day that isn't overcast. These first three hypotheses are not only falsifiable but also false, since they have already been falsified. The fourth hypothesis is not false (at least as far as we know) but would be falsified by the observation that the positive ends attract, for example. Here are some hypotheses that are not falsifiable:

> Some swans are black
> If you place a female comma butterfly in a cage with a stalk of stinging nettle,
>   it may lay eggs on it
> A rectangle has four corners

It is impossible to come up with a potential observation that would falsify the first hypothesis. Even if you only observe white swans, it is possible that a black swan may turn up one day. Likewise, the second hypothesis is not falsifiable because it is so vague that any possible outcome would be compatible with it. If a geometric figure has a different number of corners than four, it is not a rectangle, by definition. Hence, the statement is circular and not falsifiable.

You may recognise the kinds of statements made by astrologists in the second of these hypotheses. It is frequently said that horoscopes are often so vaguely formulated that the statements can be considered to be fulfilled whatever happens during the day (something along the lines of 'The signs look favourable for economic activities'). While this is often the case, we should point out that from time to time, astrologers *do* make stronger claims that clearly are falsifiable. The problem in these cases is that there is no built-in mechanism of self-correction, of learning from their mistakes. In Popperian terms, astrology is not a science partly because astrologers often don't bother to make falsifiable hypotheses and also because they don't abandon hypotheses when they are falsified.

3.2.4 Clarity and Precision

A good Popperian hypothesis thus has something to lose. It is informative because it makes a clear statement about what kinds of observations would be incompatible with it. A good hypothesis is not one that covers all possibilities but one that rules out as many possibilities as possible. This means that hypotheses can vary in their degree of falsifiability, and scientists should always strive to formulate hypotheses that are highly falsifiable, since the more potential outcomes they rule out, the more informative they are. One experiment that is often conducted in behavioural ecology classes involves the demonstration of the ideal free distribution – the idea that individuals in a population will distribute themselves among available resources so as to minimise competition or maximise individual resource acquisition. One variant of this experiment is to place students on opposing ends of a pond with ducks and instruct them to release food for the ducks. One group is instructed to release food at twice the pace as the others, thus creating a feeding patch that is twice as valuable as the other. How should the ducks in the pond distribute themselves among these two feeding patches? One could think of several potential hypotheses, for example:

1. Some ducks will go to one feeding patch, and some to the other
2. More ducks will go to the patch where more food is released
3. Twice as many ducks will go to the patch where more food is released

These hypotheses are ranked in order of falsifiability. Everything that would falsify hypothesis 1 would also falsify the two others, and every outcome that would falsify hypothesis 2 would also falsify hypothesis 3. But the reverse is not true. All hypotheses would be falsified by an outcome where none of the ducks went to feed. There are several potential outcomes that would falsify hypothesis 3 but not hypothesis 2 (such as 1.5 or 3 times as many ducks visiting the high-quality patch). Thus, all hypotheses are falsifiable, but for the falsificationist, hypothesis 3 is preferable to the others. It offers more clarity and precision precisely because it is more falsifiable.

3.2.5 Predictions

A hypothesis is a statement about the world. It can be true or false, but its information content can be quite complex, and it is often difficult to see how a hypothesis can be directly put to a test. Indeed, this can be hard even with very simple hypotheses. One such simple hypothesis could, for example, be that 'mood in humans is affected by the colour of the surroundings'. It isn't obvious how this could be tested directly, since there are many other factors that can influence mood (relationships, hunger, disease, etc.). We somehow need to be able to tell if the variation in mood is actually caused by the colour rather than any of the other potentially contributing factors. So, how did Popper mean that hypotheses can be tested, really? The answer is that hypotheses are not actually tested directly.

Testing a hypothesis is instead done by deducing, logically, what would be a *prediction* from the hypothesis, for example, under a given experimental setup, and then testing this prediction. Such deductions come in the form of *if-then* statements. For example, *if* this hypothesis is true, *then* the mood of humans should differ between differently coloured surroundings after a time spent under otherwise identical environmental conditions. As you can imagine, deducing predictions, and devising accurate conditions to test them, is not trivial. Which set of humans should be included in the experiment? How will you design the environments? How much time should be spent there, doing what? How do you measure mood? An experimental scientist will often spend considerable time designing experiments so that predictions can, in fact, be made and tested. The important point to remember, though, for now, is that the predictions are logically deduced from the hypothesis. While universal statements cannot logically be derived from singular statements, singular statements – the prediction – can be derived from general ones. Following strict Popperian falsificationism, if such a prediction is falsified, logic dictates that the hypothesis that it was deduced from must also be false. Thus, you can test and potentially falsify a hypothesis by testing its predictions.

An interesting consequence of this view of hypotheses is that for the falsificationist, a hypothesis that isn't properly tested isn't worth much; it becomes important to science only after it has been thoroughly tested and not falsified. It follows that the extraordinary carefulness that went into the formulation of hypotheses in the empiricist tradition can safely be thrown away. Scientists are free to come up with hypotheses any way they want, and since they become part of science only after rigorous testing, there are no limits to how 'crazy' they can be. The function of the scientific procedure is to eliminate false hypotheses, and the only thing that constrains the formulation of novel hypotheses is the imagination of the scientist.

3.3 Mechanisms and Change

It may be in place to say a few additional words on the claimed efficiency of hypothesis testing over empiricism, as we have noticed that it is often a source of some confusion. After all, we do use inductive reasoning in many of our everyday decisions, and it seems to work quite well for us. Why would we rely on inductive inference in everyday decisions if it were inefficient?

Consider a group of farmers who are struggling to get enough to eat from their fields. One year they notice that the yield is much higher than previous years, and of course they would like to repeat this success next year. What do they do? Our bet is that they try to remember everything they did last season and attempt to repeat all these actions as accurately as they can. If there were some actions among these that actually had a positive effect, it is likely that they will succeed again, and by inductive reasoning they conclude that the actions they performed are contributing towards a better harvest, be it the clothes they wore when ploughing or the moon phase when sowing. Each year they see their procedures work, they will get further confirmation of their conclusion.

They could instead have seen it as an opportunity to test a series of hypotheses, perhaps by doing controlled experiments, dividing all the farmers' plots into different treatments to isolate the effects of each potential factor. But they probably wouldn't, because chances are high that some (or all) of these farmers would starve as a consequence. Inductive reasoning led them to a working solution, but it wasn't very efficient as a method to gain knowledge of the true causes behind the observation. And that is typically the goal of science. Indeed, the ability to find and isolate mechanisms and causal relationships has been integral to the success of modern science.

We would go so far as to say that this is one of the main reasons for why Popper's ideas gained traction in the scientific world. In Chapter 2, Mary observed a large number of times that female peacock butterflies laid their egg clutches on the underside of nettle leaves. The empiricist may then conclude with reasonable certainty that this is a general pattern. But *why* do they show this behaviour? What are the mechanisms involved? Such questions are not easily answered through inductive reasoning but quite approachable for a falsificationist.

A related advantage is that by unravelling the mechanisms behind the patterns we see, falsificationists can make informed predictions about how the world should behave under various circumstances and test these. As a result, they are in a much better position to understand *change*. It is sometimes claimed that the statement 'the weather tomorrow will be the same as today' (which is based on inductive reasoning) is as accurate, on average, as the professional forecasts. We don't know if this statement is really true, but even if it is, it isn't very helpful, since what we are typically most interested in when it comes to weather forecasts is predicting when the weather will change. Since falsificationists are in a better position to understand mechanisms, they are also in a better position to predict change. If predictions fail, they can be modified and tested again and thus increase in accuracy over time.

In this context, there is yet another important advantage of falsificationism over empiricism, one that had an immense impact on scientific progress during the twentieth century. The shift from focusing on observable 'facts' to testing hypotheses about the world makes it more permissible to also study non-observable mechanisms. As long as one can make predictions about how these non-observable mechanisms would affect the world – and test them – they can be studied. The unobservable can be understood by predicting their effects on what is observable. By testing these predictions and adjusting our hypotheses if needed, our understanding of such processes can be further refined. This has made it possible to gain knowledge about molecular and even sub-atomic processes, as well as phenomena such as supernovas, dark matter and black holes.

3.4 Ad Hoc Hypotheses

As we have seen, Popper's ideas on falsification have some distinctive advantages over the empiricists' approach, where theory is derived from experience through careful observation. According to Popper, scientific hypotheses – and the theories they are derived from – are tentative constructs that are always subject to revision. In contrast to the empiricists, Popper was very clear that the truthfulness of a hypothesis may never be determined with any certainty. Any hypothesis is subject to falsification, and if it hasn't already been falsified, it could well be in the future. With time, this will mean that erroneous hypotheses will be pruned. Thus, in its essence, falsification is a destructive process. If there is to be scientific progress beyond the refuting of discarded hypotheses, there must also be a way to constructively modify hypotheses that have been falsified, so that they can hopefully better stand forthcoming tests. How is such refinement of hypotheses supposed to be done? How do we know when a modification is justified, as opposed to a complete falsification?

Suppose that you do have the hypothesis that 'all swans are white', and that you on one of your excursions come across a swan that by all looks of it is indeed black. Have you falsified your hypothesis? Perhaps, but it is possible to save a refined version of it. The simplest way of doing this would be to rephrase the hypothesis so that it now instead says 'all swans are white, *except that particular swan*'. There is nothing logically wrong with this, and should further observations of black swans accumulate, new versions of the hypothesis can be constructed in much the same way. In this simple example you may feel strongly that this is unsound. But why?

The reason why this is not an advisable way of revising your hypothesis is that it is only done to save your original hypothesis from falsification. By doing this, you have not really learnt anything, save for the knowledge that was inherent in the aberrant observation itself: that there is an exception. From a falsificationist standpoint, there is no way to falsify the modified hypothesis that would not also falsify the original one. It is not *independently testable*. Popper called such hypotheses ad hoc, a Latin phrase that translates as 'to this', implying a modification that is tacked onto the original one for the sole purpose of saving it from falsification. We have learnt that hypotheses should be as falsifiable as possible, and in fact, the ad hoc hypothesis is less falsifiable than the original hypothesis.

Are there proper ways to modify the hypothesis in cases like this? Yes, one can envision a number of new hypotheses that are indeed independently testable. Perhaps you have found a species of swans that are indeed black (such as the *Cygnus atratus* mentioned in Section 3.2.3)? Or perhaps the 'black swan' is in fact not a swan but a new species that looks like a swan but is not closely related to 'true' swans? There are ways of finding this out. Perhaps the observed 'black' swan was covered in oil? Again, there are ways to test this that do not involve testing the original hypothesis (the swan can be cleaned, or less advisable, a white swan can be covered in oil to see if it also becomes black). It seems clear that ad hoc hypotheses should be avoided, and from this simplistic example, it seems pretty easily done. In a more complex situation, it may not be quite as easy. If the new hypothesis is independently testable, we can instead call it a post hoc ('after this') hypothesis. This would be a modification of the original hypothesis in light of new observations, so that it is independently testable, but not actually tested (yet). In real situations, the line between the two can often be a bit blurry. After all, experimental design is difficult and it may not be immediately clear how a modified hypothesis could be tested. Especially if doing so would require not-yet-developed methods or equipment.

Another issue comes more from practice than logic. Consider another hypothetical example of what may happen during the testing of a hypothesis through an experiment, especially designed for this purpose. Imagine that it turns out that the predictions are not met. When writing up the work for publication, the scientists then come up with a new quite reasonable hypothesis that is consistent with the results. We suspect that a common way to deal with this situation is to rewrite the manuscript as if the experiment from the beginning was already a test of the new hypothesis (because it will be easier to publish this way; it makes for a better 'story'). The scientists in question could argue that it shouldn't matter much, because if they would have had the second hypothesis to start with, the outcome of the experiment would have been the same. The problem, of course, is that the new hypothesis is not actually tested. If the new hypothesis was an ad hoc modification of the original one, readers of the paper might never know. It would be slightly less problematic if the new hypothesis was independently testable (post hoc). Here, the fault was not to come up with the alternative hypothesis but to present it to the world as if it was tested, while it was in fact designed to match the outcome. To test it, the scientists would have to come up with a new experiment that tests the new hypothesis, and it should be designed so that a falsification of the new hypothesis is not also a falsification of the original one. This part could, of course, also be done by someone else, but that presupposes that the article is written so that it is clear that a new experiment is indeed needed to test it.

Science does not always proceed in an idealistic way. Even if scientists are careful in presenting an interpretation as post hoc, and thus in need of further testing, it is all too easy for such hypotheses to take on a life of their own in the literature, so that they eventually will be perceived as properly tested hypotheses. Apparently, hypothesis testing is not as straightforward as it may have seemed. In Chapter 4, we will take a deeper look at some further complications and difficulties with falsificationism.

4
Hypothesis-Driven Science: Limitations and Alternatives

Can we really falsify a hypothesis?

4.1 The Limitations of Hypothesis Testing

Popper's ideas did not go unchallenged. Some of the potential problems with falsification were noted by Popper himself, and he meant that these challenges could be fended off. Nevertheless, the problems have piled up, and we must concede that the somewhat simplistic view of falsificationism that is conveyed in Chapter 3 is overly naive. Some elaboration is needed, and perhaps some backtracking. Now, let us see how serious these challenges against falsificationism are and what could be done about them.

The first problem is rather simple but still troubling for the strict falsificationist. What constitutes a falsification, really? Is it enough with one single observation (such as a black swan) to reject a hypothesis? In reality, few hypotheses are as simplistic as 'all swans are white', and few scientific generalisations are entirely without exceptions. Moreover, an additional layer of complexity can come from the subject matter itself. While it may be reasonable to expect two gold atoms to truly be identical, it is not a reasonable expectation for some other subject matters. For example, even if you repeat a biological experiment exactly the same way, the outcome may still differ somewhat because biological entities are never identical. Similar issues exist in other fields of science, too, like in geology or many of the social sciences. We will return to this issue in more depth later in this chapter and especially in Chapter 9. For now, it is important to note that it is very rare that a single aberrant observation would be enough to falsify a hypothesis. This isn't a deathblow to falsificationism as such – it is just that we must accept that several 'falsifications' are typically needed to truly falsify a hypothesis. It does, however, introduce a troubling complication, since we now need to address the question of just how many of these we need for a conclusive falsification. It would seem that even the falsificationist would have to resort to inductive reasoning to answer this question.

4.1.1 The Duhem–Quine Problem

The next problem is perhaps more severe. It is commonly attributed to the French physicist Pierre Duhem and the American philosopher Willard Van Orman Quine, and as a consequence it typically goes under the name the Duhem–Quine thesis. The easiest way to explain this problem is to return to the issue of theory-dependence of observations. This theory-dependence implies that observations are fallible and subject to revision. We have said that this was a major problem for empiricists, since they claimed that observations should precede theory. But we also said that it was less of a problem for falsificationists, because they fully embraced the notion that observations are theory-dependent, even to the extent that observations are only interesting in the light of theory.

However, the realisation that observations are theory-dependent and fallible creates a somewhat embarrassing dilemma for the falsificationist. Say that we design an experiment to test a hypothesis. Suppose that we run the experiment and find that the outcome refutes the hypothesis. Consequently, we have falsified the hypothesis. Or have we? Since we have admitted that observations are fallible, what is there to say that it is the hypothesis that we should refute and not the observation? After all, we have stated early on in the book that observations (such as 'an oak tree') are in fact theories in themselves. The testing of a hypothesis thus requires a number of auxiliary hypotheses. Some additional examples are the theory behind the machinery used to make the observations, the theory required to interpret them correctly, and the assumptions involved in the testing situation itself (such as the test subjects being properly handled and that the correct settings are set in the machinery used to perform and interpret the results). In other words, a negative outcome of the experiment may imply that the hypothesis is wrong but could just as well imply that any of the other assumptions are wrong. There is nothing in the logic of the situation that allows us to blame the falsification on the hypothesis. We simply cannot tell. Popper taught us that a hypothesis cannot be confirmed, but it now seems that it also cannot be falsified. This is indeed troubling. And it is not just a 'philosophical' problem; any scientist will be able to confirm that experiments that actually 'succeed' in falsifying the hypothesis are difficult to interpret, let alone publish. This is because predicted results can be readily understood from current knowledge leading to the formulation of the hypothesis in the first place, but failed predictions are not as easily explained.

4.1.2 Falsification and Confirmation

Popper famously wrote that:

> I can therefore gladly admit that falsificationists like myself much prefer an attempt to solve an interesting problem by a bold conjecture, *even (and especially) if it soon turns out to be false*, to any recital of a sequence of irrelevant truisms. We prefer this because we believe that this is the way in which we can learn from our mistakes; and that in finding that our conjecture was false, we shall have learnt much about the truth, and shall have got nearer to the truth.[1]

While this claim may seem sound, we have just seen that an experiment that succeeds in falsifying our hypothesis is difficult to interpret, and the results will be hard to publish. Most researchers would tend to call such experiments unsuccessful. Instead, the results that are easy to publish are those where we *fail* to falsify our hypothesis. There is some irony in this, and it certainly seems worthwhile trying to understand why researchers seemingly not only fail to heed Popper's advice but tend to do the opposite.

Popper was firm in his belief that a confirmation (as in establishing a hypothesis as true) is not possible in science. When a hypothesis withstands repeated attempts of falsification, he preferred to say that the hypothesis had been *corroborated*, which can be read as 'tested and not (yet) falsified'. The hypothesis will have received some support by our failure to falsify it, but Popper emphasised that this support is only temporary. Subsequent tests may still overthrow it.

However, Popper also advised us that our predictions should be precise and bold; they are informative by ruling out as much as possible of the potential outcomes (recall the duck example in Chapter 3, Section 3.2.4). But by the same logic, and assuming for the time being that we can falsify these hypotheses, we will not learn much by doing so. We have merely shown that yet another wild idea turned out to be wrong. If you for some reason hypothesise that a certain metal alloy will show anti-gravitational properties, your fellow scientists would probably consider it with much suspicion. But from a Popperian perspective it is a valid hypothesis as long as it is testable. You may perform such a test, but when you falsify your bold hypothesis, it is hard to claim that you have contributed much to science. If you *failed* to falsify it though, even when seriously trying to, you may have made a revolutionary discovery. You can turn this example around, too. You may hypothesise that the alloy in question is affected by gravitation just like any other object. This time, your colleagues would consider the hypothesis sound but not very exciting. If you now perform a test and

---

[1] Conjectures and Refutations 1962, p. 231. Italics in original.

fail to falsify the hypothesis, you would receive little more than a yawn from them. But if you do falsify it, well that would be something!

It would seem then that you actually don't learn from falsifying the types of precise hypotheses that Popper advocated, but from failing to falsify them. In fact, to truly learn from falsifications, we would have to formulate our hypotheses to be as inclusive as possible, quite the opposite from what Popper advocated. This is also reflected in scientific practice. Scientific publications actually rarely present falsifications but failures to falsify. A typical scientific paper presents a hypothesis, outlines how it will be tested, and reports that the hypothesis withstood this attempt at falsification.

To save falsificationism then, we need to make some adjustments and some concessions. It is simply not true that scientists should rejoice when their hypotheses are falsified. This is both because we will not learn much by doing so, and because, strictly speaking, we cannot conclusively falsify our hypotheses anyway. This is a serious issue for the falsificationist. In practice, the problem can be diminished, for example by trying to devise alternative ways of testing the hypothesis that make use of different complementary assumptions. Hypotheses that withstand such repeated tests will receive increased support, and with time they will be moved into the current background knowledge. Even so, we are a far cry from the neat outline of falsificationism that we presented in Chapter 3.

4.1.3 Scientific Progress

In Chapter 2, we argued that a central goal of any theory of science should be to offer a compelling view of scientific progress. This is an area where falsificationism differs fundamentally from empiricism and its derivatives. If you remember, Bacon's view of progress seems to have been a gradual increase in generality of the 'axioms' that the method uncovers. In principle, this sentiment appears to have been shared by later offshoots of the empiricist tree, like the positivists, where the goal of science is the construction of a system of general statements that satisfy certain criteria, such as meaningfulness or verifiability.

What exactly would scientific progress be from a falsificationist perspective? A simplistic answer that we hinted at already in Chapter 3 – and that is inherent in the quote from Popper himself in the beginning of the previous section – would be that progress comes from removing erroneous hypotheses. Something like chipping away at a slab of rock. While this picture of progress does have a certain attractiveness, it still feels a little unsatisfactory, since science would then at its essence be a destructive process. As such, it does not really explain the growth of scientific knowledge.

It should be said, however, that Popper did stress the importance of corroboration – the failure to falsify hypotheses – especially in his later writing. Indeed, from what we learnt in the previous section, corroboration plays a much more important role in practice than falsification itself. It is the hypotheses that stand the test of time that are important, not the ones that fall by the wayside. But as we have seen, at best this corroboration comes about in an indirect way. Since we can never be sure that support of a hypothesis is warranted, there is a need to test it repeatedly and preferably using different kinds of tests. This is a tedious and costly endeavour, especially since there are no clear criteria that can tell us when we should be satisfied and consider a hypothesis to be thoroughly corroborated. Again, and somewhat ironically, the falsificationists seem to run into similar problems as the empiricists here.

Such problems notwithstanding, another way of describing scientific progress from a falsificationist point of view would be to focus on the growth of knowledge. All hypotheses are formulated and tested against the *background knowledge*. This background knowledge is ever changing and has to be understood in a historical context. The hypotheses that are interesting to pursue and spend effort and time on testing are the ones that clash with the background knowledge. If such a hypothesis withstands testing for an extended period of time, it will eventually move into the background knowledge itself. Thus, from then on it will be part of the current state of the art in the field. Scientific progress would then constitute the gradual growth and adjustment of the background knowledge as a consequence of hypothesis testing. Where the Baconian view of progress was mainly about increasing the generality of our conclusions and our degree of certainty in them, this falsificationist view focuses more explicitly on the growth and refinement of knowledge.

4.1.4 The Logic of Scientific Discovery?

In his criticism of empiricism, Popper went to some length to point out its logical shortcomings. You can never draw a logical general conclusion from a limited number of observations, no matter how many observations you make. This seemed at first to be a clear advantage of falsificationism. The logical argument itself is sound: if you can conclusively demonstrate that one observation is in refutation of a hypothesis, the hypothesis must be wrong. The problem, as we have seen, is that such conclusive demonstrations are not possible. At this point, little remains of Popper's logical advantage and his claim that we learn from our mistakes. A falsification only tells us that something is wrong; it gives us no logical guidance as to where the blame should be put. The logical case for falsificationism thus stands on rather loose ground.

Another famous claim by Popper – that falsificationism provides a solid demarcation for differentiating between science and non-science – also appears questionable. In Chapter 3, we learnt that, for example, astrology often made statements that were not falsifiable, and when it did make such statements, it lacked a framework to learn from its mistakes. It is doubtful if falsificationism fares much better in this regard. In fact, there is virtually no scientific theory that has never been 'falsified'. In most of these cases, scientists chose to place the blame on something else than the hypothesis. Indeed, Popper has himself argued that this was a rational standpoint and that a little dogmatism is needed to allow our hypotheses to reach their full potential [9]. If we give up our hypotheses too easily, we may never see them flourish. But when should we then abandon a hypothesis in the face of a falsification, and when should we stick to it? Unfortunately, falsificationism offers little guidance for this decision. It is undoubtedly true that when a hypothesis is refuted, scientists often tend to stick to it anyway, at least until something decidedly better comes along. In that sense, if there are competing theories, all refuted to some extent, we should perhaps stick to the one that is the least refuted. Of course, quantifying relative degrees of refutation is not easy and would seem to leave an undesirable room for subjective opinion. If all positions are allowed in the face of an apparent falsification – from strictly rejecting the hypothesis to stubbornly clinging to it – it would seem that almost anything goes.

This was indeed the provocative conclusion drawn by another influential philosopher of Austrian heritage: Paul Feyerabend. Like Popper, Feyerabend spent most of his career in the Anglo-Saxon world, mainly in the USA. In 1975, he published his book *Against Method*, a rather harsh confrontation with Popper and his followers [10]. According to Feyerabend, the problems outlined above showed that falsificationism had no rational justification, any more than inductivism or any other proposal of a universal scientific method. He argued that all these methods were so flexible and permissive in their application that the only reasonable conclusion one could draw was indeed that as far as science is concerned, anything goes. Most philosophers would not go this far though, and there are several attempts to save falsificationism, or science itself, from such relativistic interpretations. We will turn to some of these in Chapters 5–6.

4.2 The Place of Hypothesis Testing in Science

In spite of its shortcomings, there is no doubt that falsificationism remains an influential force in most natural sciences. One of the most important and lasting influences that Popper and his falsificationism have had on modern science is the strong emphasis on hypothesis testing. Inductivists spent much effort describing the proper way to formulate sound hypotheses with a strong empirical foundation. According to Popper, formulating a hypothesis is where the scientific process starts, not where it ends. The hypothesis needs to be tested, and it is only if it withstands such tests that it can eventually hope to become part of the scientific fabric. Anyone opening an article in a scientific journal can see this influence reflected in the formalised structure of the articles. The hypothesis is not something arrived at towards the end of the Discussion, but it is expressed at the onset, typically at the end of the Introduction. The role of the introduction is to provide an expression of the background knowledge, and an argument for why the hypothesis is worth testing within this context. The Methods section describes in some detail how this test is done, and the Results section presents the outcome. Finally, the Discussion provides an interpretation of the outcome in the theoretical context outlined in the Introduction and suggests how the background knowledge should be modified in light of the new results. The notion that hypotheses should be severely tested has indeed been influential, but some would argue that falsificationism itself has been too influential. Not everyone agrees that falsification is the best way to go about testing your hypotheses.

4.2.1 Taking a Step Back

As we have seen, aside from emphasising the need to test hypotheses, Popper also maintained that any modification to a hypothesis after the test must be independently testable (i.e. not ad hoc). Modifying a hypothesis just to match new evidence is circular and should be avoided.

There is an increasingly influential line of thought that has challenged this stance. To illustrate this way of thinking, let us return to the example we gave in Chapter 3 to demonstrate an advantage of falsificationism over inductivism. We used the analogy of weather forecasts to claim that induction alone is poor at predicting change, and because falsificationists emphasise making and testing predictions from presumed mechanisms, it should be better equipped to understand change. We must now admit that there may be more to this story. Say that certain cloud formations that you observe on Wednesday make you predict that there will be rain on Saturday. This is a reasonable and falsifiable hypothesis that can be tested, but what would the best course of action be? Should you stay inside the rest of the week and then come Saturday walk out with an umbrella, perhaps to see that your prediction was falsified by blue skies and sun? Hardly. Well before it was time to go out, you would have better information available than you had when you initially made the prediction.

Clearly, in this situation, it would be better to keep looking out the window and *update your predictions* as you gather more information. Using new information as it becomes available should improve your hypothesis, so why would we refuse to use it just because it isn't independently testable?

This is a point of view taken by Bayesianism, an analytical approach that has become popular across many fields of science and that has also been proposed as a philosophy of science in its own right, which avoids many of the shortcomings we have seen troubling the empiricist and falsificationist approaches. We will take a closer look at this approach in the final part of this chapter to see if it can help us solve the problems that have accumulated thus far.

4.2.2 Bayesianism

The Bayesian approach is based on the so-called Bayes' theorem, developed by the English eighteenth-century statistician and philosopher Thomas Bayes. In a nutshell, the theorem describes how probabilities – or degrees of belief in a hypothesis – should change in the face of new evidence. If a falsificationist is presented with evidence that is in conflict with their hypothesis, they should consider it falsified, formulate a new hypothesis, subject that to a new test, and so on. According to the Bayesians, however, it is wrong to treat hypotheses in this dichotomous fashion. Some theories are relatively well established, while others are more dubious, and importantly, our belief in them will change as we encounter new evidence. This degree of belief, they argue, is better expressed as a probability that can take any value between 0 and 1, where 0 means impossible and 1 means certainty.

To take an everyday example, let's pretend you are a child the day before your birthday. You find a package with your name on it, and you want to know what is in it. You had three things on your wish list: a sweater, a book and a LEGO kit, and you hope it is the latter. You know your parents only give presents that are on your list, so it must be one of these three items. What is the likelihood that the package contains the LEGO kit? Before even looking at the package, the likelihood should be 1/3, or about 33%. Seeing that the package is rather large is enough to adjust these probabilities, because a book would not require that large a package (although you cannot rule it out entirely; your parents have been known to trick you with packaging before, so you shouldn't reduce the likelihood to zero). You then proceed to pick it up and notice that it has a substantial weight, although for its size, it is not heavy. Again, this makes you adjust the likelihoods in favour of the LEGO kit, this time primarily against the sweater. Finally, you shake the package and notice an unmistakable rattle. At this point, your likelihoods have been adjusted so that you are close to 100% certain that it is the LEGO kit.

The rationale for this move from a likelihood of 33% to close to 100% is expressed in Bayes' theorem. To understand the logic behind it, we need to introduce some terminology. The general belief we have in a hypothesis at the onset can be called the *prior probability* – simply how strong our belief is before the new information. If presented with new evidence, our beliefs will need to be adjusted. The adjusted probability in the face of this new evidence is called the *posterior probability* ('after the fact'). The difference between these two probabilities is largely determined by how surprising the evidence is in light of what we already thought we knew.

To calculate the posterior probability, we need to know two additional probabilities, in addition to the prior. Firstly, the likelihood of the new evidence regardless of our hypothesis. The reason for this is that if the new observation would be expected also for reasons that have nothing to do with our hypothesis, it has little bearing on the hypothesis. It would support many other hypotheses as well and should consequently not have a strong impact on our belief in this particular hypothesis. In the birthday gift example, the size of the package adjusted our beliefs somewhat, but not very much, because many things could fit in a package of that size (especially if your parents may trick you). Secondly, we need the probability of the new evidence, given that our hypothesis is true. Like the posterior probability, this is a *conditional probability*, and it expresses how strongly the evidence is predicted by our hypothesis. Again, if this connection is weak, the evidence should not influence our beliefs so much. This is why the rattling sound of your birthday package when you shook it made you almost certain that it contained a LEGO kit. LEGO kits invariably give off a rattling sound when shaken, and books and sweaters do not. Thus, the rattling sound is strongly predicted by the hypothesis that it is a LEGO kit. By putting these probabilities together, we can now create a general formula for how beliefs should change in the face of evidence (the vertical bar '|' below can be read as 'given the'):

$$\text{Probability}(\text{Hypothesis} \mid \text{Evidence}) = \text{Probability}(\text{Hypothesis}) \times \text{Probability}(\text{Evidence} \mid \text{Hypothesis}) / \text{Probability}(\text{Evidence})$$

This is Bayes' theorem. Usually, it is given in a more abbreviated form. For example, using the first letters from the equation above as abbreviations, we get

$$P(H \mid E) = P(H) \, P(E \mid H) / P(E)$$

Note that the prior probability $P(H)$ – that is, how much we believed in our hypothesis at the onset – is in the numerator. This means that the stronger we believe in something, the more difficult it will be for any evidence to influence our belief. This may seem undesirable, since it would mean that you can cheat the calculation simply by believing so strongly in your pet hypothesis that hardly anything will make you drop it. However, it is easy to see that this makes sense. For one thing, we have all met people with such strong convictions that no evidence whatsoever will seem to change their minds, and Bayes' theorem gives an explanation for why it is so hard to sway strong believers with any argument. More importantly, it also neatly explains why very well-supported hypotheses (gravitation, evolution, ...) would require exceptionally strong counter-evidence for researchers to consider abandoning them. In a sense, this solves the dilemma that falsificationists were left with in Section 4.1.4, when Popper claimed that it is sometimes important to be dogmatic so that we do not let hypotheses go too easily. Bayes' theorem explains when a little dogmatism is rational and when it is not.

Interestingly, Bayesianism offers up a very different kind of solution than falsificationism to the raven paradox that we introduced in Section 2.4.1. If you recall, the empiricist approaches could not escape the paradox that stated that, logically, observing a green umbrella would offer just as much confirmation for the hypothesis *all ravens are black* as actually observing a black raven (because the hypothesis is logically equivalent to its contrapositive: *all non-black things are non-ravens*). The falsificationist escapes the paradox by instead looking for non-black ravens to *refute* the hypothesis. Bayesianism, on the other hand, argues that it is, in fact, not a paradox at all. If we divide the world into the two categories of black ravens and non-black non-ravens, each observation that falls into any of these categories *does* support the hypothesis. However, since there are such an enormous number of non-black non-ravens in the world, such observations would only strengthen our beliefs to an infinitesimally small degree. In other words, Bayesianism offers an explanation for why our intuitions clashed with simple logic in the first place.

Bayes' theorem can also be used to compare hypotheses by testing which of a number of alternative hypotheses is best supported by available data. In subjects that study complex phenomena, such as ecology and medicine, this approach has become increasingly popular. The usability of Bayesian statistics cannot be denied. There are many situations where you need to calculate the probability of events in light of other events, evaluate the support of alternative hypotheses or estimate how beliefs should change given some new information. Applications outside pure statistics include gambling, bioinformatics, epidemiology, machine learning and weather forecasts. However, some philosophers mean that the theorem can also inform general philosophy of science in a more fundamental way than what has been expressed so far [11].

4.2.3 Bayesianism as a Philosophy of Science

The basic problem that Thomas Bayes set out to solve was as follows [12]:

> *Given* the number of times in which an unknown event has happened and failed: *Required* the chance that the probability of its happening in a single trial lies somewhere between any two degrees of probability that can be named.

At its foundation then, Bayes' theorem is a way to calculate the probability of a future event based on how often it has happened in the past. This may sound familiar to you. Isn't it the exact same problem that empiricists like Bacon tried to solve through the inductive method? One way of thinking of Bayesianism then, as a philosophical approach, is that it is a more sophisticated method of making inductive inference. Yet, Bayesianism is an odd bird in this respect. We have treated the Bayesian approach here in relation to hypothesis-driven science, since evaluation of hypotheses in the light of new evidence is at its very core, and it is often used in a hypothesis testing framework. In some respects, its rise in popularity can also be seen as a response to the perceived shortcomings with the Popperian style of hypothesis testing that we have discussed in this chapter. Why do we feel that we have learnt something when we falsify a bold hypothesis but not when we falsify a cautious one? Why do scientists sometimes cling to their hypotheses in the face of apparent falsification? The Bayesian machinery claims to offer straightforward answers to such questions. It would appear then that Bayesianism might bridge the gap between what we have called observation- and hypothesis-driven approaches to science.

Unfortunately, the Bayesian approach is not without its own problems. Bayesians do make some fundamental claims, such as that degrees of belief can be expressed as probabilities in a meaningful way and that these degrees of belief obey the axioms of the probability calculus. For the most part, these claims seem reasonable. For example, if your belief that it will rain tomorrow is 0.7, your belief that it will be overcast should not be 0.3, as it should not rain if it is not also overcast. Such a set of beliefs would be irrational and can be shown to produce unrealistic outcomes. In betting, it can lead to so-called *Dutch books*, where you can be made to lose no matter the outcome. However, a first problem with this is that human beings are not omniscient. It is quite possible (and common) to hold beliefs that are logically inconsistent, even for scientists. Moreover, is it reasonable to assume that you can really assign probabilities to all scientific theories in a meaningful way? What is the probability of Newton's law of gravitation, for example? What does such a probability even mean? And where does this probability come from?

Indeed, the issue most often mentioned with regard to Bayesianism is how to get the Bayesian machinery started in the first place. Again, where do the prior probabilities come from? In some cases, such as with the birthday gift example given in Section 4.2.2, or when drawing a card from a well-shuffled deck, the priors are relatively easily given, because there is a finite number of potential outcomes and thus a finite number of potential hypotheses to choose from. There were three possible gifts in our example, and a deck of cards contains 52 cards. In the absence of any other information, the prior probability should then be 1/3 and 1/52, respectively. However, scientific problems are rarely like this. We typically don't know how many potential explanations there are to our scientific problem, and theoretically, the number of potential hypotheses should be infinite. If so, the prior probability of any hypothesis should be 1 divided by infinity, and the calculation cannot even get started. Falsificationists do not have this problem since they don't try to assign probabilities to theories. Hypotheses can only be untested or tested and falsified or not (yet) falsified.

Bayesians differ among themselves on how to solve this problem of the priors. On one side, we have the *objective Bayesians* who argue that priors should be set by assigning the probabilities that a rational agent should assign, given the information available [13]. If no information is available, one should assume equal probabilities, like we did with the birthday gift (Section 4.2.2) and the deck of cards above. As we just said, this may run into problems if the number of possible explanations is potentially infinite. However, it should be acknowledged firstly that there are indeed some valid scientific problems where there really is a finite number of potential alternative hypotheses to choose from. Moreover, even when there is not, it can be instructive to ask which out of a finite set of proposed hypotheses is best supported by current evidence. In many scientific applications, this is in fact how priors are set, and as long as conclusions do not expand beyond these specific hypotheses, this should be perfectly fine. With this caveat, this means that the objective approach to assigning priors should be usable as a scientific method of analysis, even if it does seem questionable as a general philosophical approach.

On the other side, we have the *subjective Bayesians* [14]. They escape the problem of an infinite set of potential hypotheses by simply stating that priors should represent the degree of belief in a given hypothesis that a scientist actually holds. While it may seem strange to introduce subjectivity into an otherwise objective probability calculus, this seems to be where most Bayesian philosophers of science have ended up. The approach also does have its appeal. It is clearly true that some hypotheses have stronger support in the scientific community than others, and subjective Bayesians can take this into account. If evidence has been piling up over the years in favour of a hypothesis, it seems to make sense to treat it as such, when, for example, comparing it with a newly formulated and untested hypothesis. On the other hand, scientists typically differ in the degree of belief they have in a hypothesis, and which belief should we base our prior on? And how exactly do you go about determining the degree of belief in a hypothesis that a scientist has (or had)? Perhaps we should set the priors according to the consensus in the field, which seems akin to articulating the 'background knowledge' in a more formal way? But our experience tells us that even such a consensus would be hard to agree upon in many cases. Moreover, to the extent that a consensus can be found, what exactly should the prior probability be set to? For example, what probability should we assign to a well-supported hypothesis? 0.8? 0.9? Who can really say? A subjective Bayesian may answer that it actually doesn't matter that much, since posterior probabilities should tend to converge anyway, given the new evidence [15]. If you believe more strongly in a hypothesis than I do, and we are presented with convincing evidence in favour of it, my beliefs should change more as a consequence than yours, and as more evidence comes in, we should eventually end up in the same place.

In this sense, the Bayesian approach can be seen as a 'consensus machine' that explains how diverging opinions should converge in the face of evidence. This is indeed an appealing feature that shines a light on an important aspect of the scientific process. This can perhaps also provide a hint into a Bayesian view of scientific progress. Scientific advances occur through an iterative process where prior beliefs are updated in light of new evidence. This allows scientists to refine their hypotheses so that they provide a progressively more accurate understanding of the subject matter.

However, there are some additional problems with Bayesianism as a philosophy of science that, to our minds, are even more troubling than the problem of the priors. First, as we mentioned earlier in this section, there is no doubt that the Bayesian approach to statistics has a number of useful applications, within and outside of science. Indeed, some of the most successful applications of Bayes' theorem come from outside of science, such as in betting. From the perspective of philosophy of science, this is a problem in itself, for in what sense can it then be said to be a philosophy of *science*?

Moreover, if scientists are truly following the Bayesian consensus machine, why do we so often have long-standing scientific controversies when we all have access to the same evidence? Shouldn't we all have converged on the hypothesis that is best supported by available evidence a long time ago? We have learnt over and over in this book that 'evidence' is not unproblematically given and that observations are always theory-dependent. If researchers differ in how much belief they have in a hypothesis, why would they not also differ in how much belief they place in the evidence? In this respect, Bayesianism does not fare much better than falsificationism or classical empiricism for that matter. In a sense, it fares even worse. As you remember, we claimed that the Bayesians could get around the problem of subjective priors by arguing that the exact value of the prior is not crucial, since diverging opinions about the prior would tend to converge in the face of evidence. If we now have to accept that it is possible to also have variable subjective beliefs in the evidence, it puts the Bayesians in a rather awkward position. If degrees of belief can vary both in the hypothesis and in the evidence, it is possible to end up in situations where one scientist could draw the rational conclusion that the evidence strengthens a given hypothesis, while another researcher could draw the equally rational conclusion that it does not. The consensus machine would never converge and we seem to be back on square one. Like the falsificationist and the empiricist, Bayesians are left with the troubling task of working out what in the complex web of assumptions we should trust, and there is as little in the Bayesian philosophy as in the other approaches we have met to assist with this task.

It appears that for any hypothesis testing to work, researchers need a framework that can guide them through this complex maze of assumptions and allow them to decide what to trust and what to reject. We will now turn to some attempts to describe how such guiding theoretical frameworks might look and function.

5
Paradigm-Driven Science

Science functions within structures of theories, where some theories are fundamental.

5.1 Mary's Butterflies

Back to Mary again!

A few years have passed since the previous adventures in Mary's garden. Her associations with other butterfly enthusiasts have matured to the point that they have formed an entomological society, and even publish a modest scientific journal – *The Insect Gazette*. Mary and most of her colleagues are quite influenced by the ideas of Charles Darwin, and the journal thus contains not only observations of insects, their traits and behaviours, but also hypotheses as to the evolutionary causes behind these traits.

Mary is no longer content with observing her butterflies and contributing to the known facts about them, or even with testing these supposed facts with experiments. She wants to understand *why* they behave as they do, and she is very excited with the possible explanation provided by Darwin: they possess the traits that have historically been favoured by natural selection – evolutionary adaptations to the environment.

Mary has even gotten to the point where she has been testing a specific hypothesis regarding butterfly host plant preferences as adaptations. It made sense to her that butterfly females should prefer to lay their eggs on the plants which provide the best nourishment for their offspring, because this behaviour must surely be favoured by natural selection. Females putting eggs on poor hosts would produce offspring that grow poorly or even die before they could themselves reproduce, and this unfit trait would thus be weeded out by selection. Unbeknownst to Mary, this exact idea has in fact already been suggested by Darwin in his unpublished writings as an example of the workings of natural selection [16].

It is now autumn, and over the summer that just has passed Mary has experimented with offspring of the peacock and the comma butterfly. You may remember that the peacock seems to be highly specialised in its choice of host plants, only using stinging nettle as food for its larvae. This limited Mary's experimental design, but she did try to move some of the small larvae to cuttings from other plants in her garden. Invariably the larvae refused to eat more than a bite, if even that, and tried to escape the plant. Mary soon took pity on the larvae and moved them back to the nettle that they came from.

The comma butterfly had presented a more interesting challenge, in that she had earlier found that females of this species sometimes accept other plants as hosts for their offspring, such as black currant. However, since they do seem to prefer stinging nettle Mary hypothesised that this is because larvae grow and survive better on the latter plant. Much to her satisfaction this was also exactly what she found when performing the experiment of putting a few larvae on each of the two plants, following their development through the summer. Most of the larvae survived on both hosts, but the ones reared on nettle reached the pupal stage more than a week earlier. This success in predicting the experimental outcome had even more deepened Mary's belief in Darwin's ideas, and her presentation of the results to the entomological society a few days ago had been much appreciated.

Today, Mary has a guest in her garden. It is the vicar of the parish, who shares her interest in insects and natural history in general. They have a nice talk over tea, sharing and comparing their butterfly observations. Mary briefly considers inviting the vicar to the entomological society, but quickly abandons the idea. Surely it would only hamper discussions, if they had to make an effort to avoid the topic of Darwinian explanations for insect traits? No, best leave talks with the vicar to a shared admiration of God's creatures.

5.2 The Philosophy of Paradigms: Thomas Kuhn

5.2.1 Normal Science and Revolutions

In the story in Section 5.1, Mary tests her specific hypothesis predicting that butterfly females (or indeed any plant-feeding insect) should prefer to lay their eggs on plants that provide the best food for their offspring. This was also one of Darwin's early ideas regarding how natural selection ought to work; after all, females with heritable properties helping them to make the 'right' choice should leave more successful offspring and thus increase in frequency in the population, relative to other variants. In modern writings Darwin's (and Mary's) hypothesis is typically referred to as the preference–performance correlation hypothesis.

What would have happened if Mary's prediction had not been fulfilled? Or if we perform a similar experiment today and do not see a positive correlation between female preference and offspring performance? As we discussed in Chapter 3, a naive Popperian view would be that the hypothesis in that case has been falsified. Taken to the extreme, this would also falsify the theory of natural selection (which it is derived from) and perhaps also the theory of evolution, since this was and is the basis for the hypothesis. In fact, the preference–performance correlation hypothesis often holds true, but by no means always. This could be for a number of reasons, including that females may not always be very choosy in their selection of host plants but instead prioritise getting many eggs out – as fast as possible. So the hypothesis might survive apparent falsification if it is still useful in many cases. And even if it would be abandoned by scientists in the field because there are too many exceptions to the rule, this would not mean abandoning the theory of natural selection, let alone the fundamental theory of evolution. Rather, new hypotheses based on evolution by natural selection would replace the preference–performance correlation hypothesis.

The idea of fundamental theories that are protected from immediate falsification is at the heart of paradigm-driven philosophies of science. It was most famously formulated as an explicit philosophy by Thomas Kuhn in his book *The Structure of Scientific Revolutions* [17]. Here, Kuhn outlined a view of science where researchers in a given field of science normally work within a single *paradigm* of accepted theoretical assumptions and research methods (following an initial disorganised *pre-science* phase). They will attempt to investigate and understand the world based on this paradigm, performing experiments based on its fundamental assumptions. The paradigm can be developed and modified over time to explain and accommodate experimental results, solving perceived problems and improving the fit between the paradigm and the real world. Kuhn termed this phase *normal science*. However, it could be that experimental outcomes too often or too seriously do not follow predictions. Results that do not lend support to the paradigm or even oppose it thus accumulate, so that researchers in the field in the end lose confidence in the paradigm itself. Such a state of *crisis* can ultimately lead to a switch to a new paradigm, a *scientific revolution*. And the cycle starts over with a new period of normal science, based on the new paradigm.

Kuhn made the point that we have also tried to make in Chapter 4, that single theories are not really absolutely confirmed or clearly falsified during the process of normal science. We can never be entirely sure that our observations are correct, since they are based on a network of theoretical assumptions, and thus they can neither be used to reliably confirm nor refute a theory. Rather, Kuhn wrote in the Introduction to his 1970 book:

> Competition between segments of the scientific community is the only historical process that ever actually results in the rejection of one previously accepted theory or in the adoption of another.[1]

---

[1] The Structure of Scientific Revolutions, p. 8.

In other words, only when an old paradigm is replaced with a new one can there be a fundamental change in scientific thinking. Single theories exist within a whole structure of theories, but when the structure itself is rejected after a scientific crisis, there is real change. A major scientific discovery can also spark such a revolution, according to Kuhn, because it not only adds to knowledge but also forces scientists to reconsider their earlier thinking and change their methods. Consider, for instance, the discovery of a major chemical element such as oxygen, with its many roles in chemical processes, and how this must have profoundly affected chemistry as performed at the time. Or the discovery of DNA as the vehicle for genetic inheritance and biological replication, and how this finding led to the huge and rewarding field of molecular biology, with its dedicated theory apparatus and specific methodology.

5.2.2 Incommensurability

Kuhn stressed that rival paradigms are based on very different views of the world; they are *incommensurable* in his terminology. He pointed out that a scientific community in an emerging scientific field would soon try to agree on fundamentals such as:

> What are the fundamental entities of which the universe is composed? How do these interact with each other and with the senses? What questions may legitimately be asked about such entities and what techniques employed in seeking solutions?[2]

In a mature science, the answers to these questions would be an important part of the research education and would consequently come to 'exert a deep hold on the scientific mind', in Kuhn's words. This provides for efficiency in research and gives direction to the scientific field but also means that science in a sense will be forced into 'the conceptual boxes supplied by professional education'. A paradigm shift, then, means abandoning these boxes and opening up to a new view of the world. This will often mean, according to Kuhn, that much of the previous research and its findings are now seen by scientists working in the new paradigm as speculative, irrelevant or entirely unscientific. Conversely, they will instead be studying many phenomena that were previously either not observed at all or explained away.

In one of his strongest statements, Kuhn wrote that it is because of the fact that scientists in different paradigms see and describe the world so differently

---

[2] The Structure of Scientific Revolutions, pp. 4–5.

(and thus cannot communicate well with each other) that we see scientific 'revolutions':

> Just because it is a transition between incommensurables, the transition between competing paradigms cannot be made a step at a time, forced by logic and neutral experience. [...] it must occur all at once (though not necessarily in an instant) or not at all.[3]

Mary's experience with the vicar provides a possible example of different paradigms if she is correct in believing that they would not have had very fruitful discussions in the entomological society with the vicar being present. Kuhn's own examples were drawn mostly from physics (his own earlier field of study), such as Newtonian physics or the Copernican Revolution, but we would propose that the rise of Darwinism and evolutionary thinking is one of the best examples of something like a paradigm shift in science. Surely a creationist and an evolutionary biologist see the living world very differently, at least when it comes to proposing explanations for the patterns observed in nature.

5.3 The Limitations of Paradigms

Kuhn's concept of paradigms has been very influential in science (more on this in Section 5.4) but there are some possible limitations that need to be pointed out.

5.3.1 Where Are the Revolutions?

One problem with Kuhn's idea is that it is not clear whether actual scientific *revolutions* really happen at all. The notion has certainly been pervasive in the history of science, where we routinely talk about the Copernican Revolution, the Darwinian Revolution and so on. It is true that these examples represent major shifts in scientific thinking with huge ramifications on their respective disciplines. Nevertheless, it can be questioned to what extent even these well-known examples represent 'revolutions'. After all, the shift to a Copernican sun-centric view took hundreds of years to complete. For any fundamental historical shift in scientific thinking, it is possible to find stragglers who never accepted the new paradigm. Kuhn was in fact well aware that not all scientists would switch to a new paradigm, citing the rise of Darwinism as an example. Thus, his insistence that such transitions must 'occur all at once' is somewhat puzzling, unless it is recognised that this should be seen as a description of what happens in the mind of an individual scientist rather than in an entire field

---

[3] The Structure of Scientific Revolutions, p. 150.

of research. Nevertheless, we would contend that describing paradigm shifts as 'revolutions', with its strong connotations of sudden change, may be somewhat misleading.

Another problem is how progress in science can ever be achieved if concepts and ideas in a field of research change so completely as a result of paradigm shifts that communication between followers of the old and new paradigms is impossible. Do scientists, then, have to start over from scratch after each revolution? And if so, how can any of the scientific knowledge accumulated in a previous paradigm be preserved and transferred to the new paradigm, ensuring progress over time?

It can be questioned if it is ever really true that followers of old and new paradigms see the world so differently that they can't even discuss natural phenomena. The Copernican Revolution eventually took place because the sun-centric view was a better explanation for observations of the natural world that all competent astronomers already agreed upon, not because a new type of observation was suddenly made. What about the rise of Darwinism?

5.3.2 The Preservation of Knowledge

Long before Darwin's ideas impacted the world, there were scientists studying the natural history of organisms and how their traits seem to be wondrously adapted to the habitat – and to each other, as for instance in the case of flowers and their pollinators. Some of them worked in an old tradition of *natural theology*, where the existence of such adaptations would be seen as evidence for the workings of a creative God. We could imagine the vicar introduced in Section 5.1 as belonging to this tradition. In contrast, Mary and her colleagues in the entomological society have aligned themselves with Darwin's ideas, meaning that they would see the exact same traits instead as the workings of evolution by natural selection.

Building on this example, biologists at least since Aristotle have noted the striking similarities between, for instance, the legs of a horse and a cow, or even a dog (Figure 5.1). The term *homologies* was later coined for such traits that share form and function, as well as relative positions in the body. In the absence of evolutionary thinking, they could be explained as different manifestations of the same particular idea in the Creator's mind or as evidence that nature favours particular archetypical morphologies. Since Darwin, however, homologous traits are explained by them being inherited from a common evolutionary ancestor. The very same phenomena in the world of living organisms are thus viewed very differently in the old and modern paradigms, in terms of explanations. For an evolutionary biologist of today, it is even hard to

Figure 5.1 Examples of homologous traits in (from left to right) horse, cow, sheep, pig and dog. Corresponding bones, groups of bones and tissues in different species are shown in the same grey tone. The hooves and claws (in black) are homologous to our nails. Homology is explained very differently before and after Darwin, exemplifying a paradigm shift.

understand how they were explained previously – a good example of Kuhn's incommensurability.

However, followers of both paradigms would still agree on many aspects regarding what exactly is out there to *be* explained: the species, their traits, their adaptations and their homologies. They do share the same world, meaning that knowledge accumulated before Darwin could not only be preserved but even be used by him to argue in favour of his idea. In order for Darwin to succeed in convincing proponents of the old view, he had to appeal to shared convictions – observations, established knowledge – and show how such phenomena make better sense in the light of evolution. Darwin's *Origin of Species* was written as a long rational argument to convince a reader immersed in natural theology. The rivalling views were incommensurable only if one refused to accept rational argument.

Kuhn's emphasis on paradigm shifts as revolutions in thinking and on incommensurability between old and new paradigms can thus be challenged on empirical grounds. To the extent that such revolutions truly happen, they can be further challenged on the grounds that they open up for relativism, the notion that science is just a matter of opinion. Kuhn himself rejected this charge, but at the very least there is a tension in his theory with regard to this. It appears that Kuhn himself not only rejected the claim of relativism but that he also saw himself as a realist, that is, he believed in the existence of an objective reality that constrains our theories, and that the role of the scientist is to refine theories by testing them against the real world. He maintained that the new paradigm's ability to solve the problems that led to the crisis of the old one shows how scientific progress can happen, even between paradigms. However, much of his writing invites a different interpretation. Sentences like 'The competition between paradigms is not the sort of battle that can be resolved by proofs'[4] imply that the perceived advantage of the new paradigm is not necessarily rational. If, as Kuhn maintained, scientists before and after a scientific revolution really live in different worlds, it follows that a paradigm-independent reality would be impossible to study and learn about. The role of the paradigm is to provide a coherent framework where productive scientific work can be done, but given this emphasis on living in different worlds, the paradigm could also be interpreted as a bespoke 'reality' that scientists construct to fit their current theories. Concepts of truth and reality thus become paradigm-dependent. If so, how can we say that one paradigm is qualitatively better than another? How can there be scientific progress?

5.4 The Place of Paradigms in Science

There is certainly much merit to Kuhn's description of scientific thinking in a particular field of research as a network of theories, with some being so fundamental that scientists in the field have to provisionally accept them as true – in order to provide a basis for all those theories that are instead explored, tested and improved within the paradigm. It is also clear that shifts in such theoretical fundamentals might result in communication problems, in particular when it comes to which type of questions would be found of interest to pursue.

Not surprisingly, then, the concept of paradigm shifts can today be found everywhere, not only in scientific writing but also as a part of everyday language. A web search on the term can turn up anything from 'a new paradigm for environmental sociology' to describing Jack Antonoff as 'something of a

---

[4] *The Structure of Scientific Revolutions*, p. 148.

new paradigm as a pop producer'. It is thus often used in a less strict manner than Kuhn intended in order to depict more or less any new way of thinking about a subject – big or small. In science this can lead to a new field of research, and on occasion, even to a new subdiscipline. In this way, recognised paradigms in science can sometimes be equally well described as *research programmes*, in the terminology of another philosopher of science that pursued a similar line of thought – Imre Lakatos.

5.4.1 Research Programmes

Lakatos (1922–1974) was a Hungarian philosopher, much influenced by both Popper and Kuhn but attempting to address the difficulties that he saw with both falsificationism and paradigms [18, 19]. Along with Popper, Lakatos recognised that we can't prove in science that theories about the world are true; we can only test them as well as we can and keep them as provisional knowledge if they stand up to such tests. Along with Kuhn, Lakatos also stressed that science functions within a larger structure of theories. In order to be able to formulate and test hypotheses, we need to have an unquestioned fundamental theoretical basis for predictions – what Kuhn called the paradigm and Lakatos called the *hard core* of a research programme. If you as a scientist start to question the hard core you leave the research programme, similarly to Kuhn's paradigm shift in the mind of an individual scientist. Straightforward examples would be that an evolutionary biologist needs to believe in evolution, and a particle physicist needs to believe in not only the existence of subatomic particles but also that they are important to characterise and understand.

In contrast to Kuhn, however, Lakatos meant that competing research programmes can exist side by side. Some research programmes may be *progressive* in the sense that they consistently come up with predictions that are supported when tested, attracting more scientists into their fold. Other programmes may be *regressive*, in that they fail to produce such new knowledge and would eventually fade away as more and more scientists leave for other, more fruitful, alternatives. Research programmes can also be hierarchical (e.g. scientific disciplines, subdisciplines and smaller fields of research) and scientists will frequently see themselves as belonging to several research programmes at once. The authors of this book, for instance, could be described as belonging to the research programmes of ecology, evolutionary biology or evolutionary ecology, to name just a few possibilities. There is a certain correspondence between research programmes and the names of scientific journals or international conferences, depicting a subdiscipline with a large enough community of researchers to have a productive dialogue – based on a shared hard core.

Lakatos developed the concept of research programmes further into a useful model and terminology for how science works. Besides the fundamental theoretical hard core a research programme at any given time has a set of general theories that are provisionally accepted as possibly true in the field and can thus be used to formulate new, more specific and more testable hypotheses. If such tests repeatedly fail, these theories will eventually be replaced with new ones, but the hard core can be preserved. For this reason, Lakatos termed this set of theories the *protective belt*. Researchers within a programme, say evolutionary biologists or particle physicists, do not always agree on these general theories within their field but are engaged in a discussion about them. What is the relative role of random versus adaptive processes in evolution? Is the hypothetical graviton particle really likely to exist? Hypothesis testing is used to attempt to resolve such discussions.

There is also a set of prescriptions for how science should preferably be performed within the research programme, the methodology do's or don'ts, or what Lakatos called the *positive and negative heuristics*. These prescriptions can profitably be seen as theories regarding how science is best done, because the methods of a research programme can evolve over time as they are tested in practice. These heuristics can also provide a solution to the Duhem–Quine problem outlined in Chapter 4, in that they can provide guidelines for what researchers should do when facing an apparent falsification. The most obvious guideline would be that such falsifications should be deflected to the protective belt rather than the hard core. Importantly, the positive and negative heuristics need not necessarily be spelled out; they are often tacit and learned by socialisation into a discipline.

There is clearly a strong similarity to Kuhn's description of normal science between revolutions, when scientists who accept the paradigm strive to find the best ways to fit theories with the world, tinkering with the theories (Lakatos' protective belt) and the methods (Lakatos' heuristics). However, Lakatos' scheme of research programmes meets some of the criticism that has been levelled against Kuhn's paradigms. In our opinion, it better describes how science actually has worked historically, doing away with revolutions and providing a way for scientific progress to be achieved. Since research programmes can exist side by side and even overlap, there can well be communication between them; they are not necessarily incommensurable. After all, there is constant communication even between different scientific disciplines such as chemistry and biology, to the extent that this is needed. As a consequence, knowledge that has been accumulated in a research programme can be preserved even if the programme itself eventually decays and goes extinct. Indeed, an emergent competing research programme should not only be able to make novel predictions that open up for new and different kinds of knowledge, but it should also be able to explain observations from the previous program.

5.4.2 The New Experimentalism

Finally, as mentioned earlier in the book, there have been more recent developments in the philosophy of science attempting to take an even further step back from the rather extreme theory-dependence in Kuhn's paradigms. This comes in the form of the 'new experimentalism' promoted in particular by Deborah Mayo [5], which suggests that there are experimental results (and associated knowledge) that stand even if the theories that explain them change, and would thus survive a paradigm shift. Mayo makes a distinction between 'low-level theory' and 'high-level theory'. Low-level theory is the theory behind the data and observations, for instance, theory regarding how an instrument works and when it can be relied upon to give accurate observations, or theory behind statistical tests. High-level theory is the scientific theory that can explain or predict the observations.

Mayo acknowledges that we can never be completely certain about the validity of our experimental results, but she outlines in detail how in practice experimenters have methods for making them reliable enough to pass as real knowledge; trained experimenters know about potential error sources, how to control for them and how to test results with statistics. Additional experiments or other kinds of tests can also be added if necessary to rule out error sources, including confounding variables. If in the end we still find with reasonable certainty that the observation is not in error, it has passed a 'severe test' and can be added to what we know about the world. In so doing, we may also be able to corroborate (or not) a hypothesis from high-level theory that could predict the observation. At a later point in time scientists may find alternative and better explanations for the experimental results, perhaps stemming from a new paradigm or research programme, but the results themselves still stand if good scientific practice was followed when they were established. In this way not only directly observable phenomena such as the homologous bones mentioned in Section 5.3.2 (see also Figure 5.1) or the lighting of a lamp in a closed electrical circuit mentioned in Chapter 2 can remain as established 'facts' after a shift in high-level theory, but also experimental results highly dependent on instruments, mathematics and other forms of theory. This constitutes *experimental knowledge* in Mayo's terminology. There are some clear connections between Mayo's ideas and those of Kuhn and Lakatos, but where the latter two focus on the development of the high-level theoretical structures that guide scientific work, Mayo emphasises the value of the actual scientific findings themselves. Much like the Darwinian and pre-Darwinian researchers in the example earlier in this chapter, scientists adhering to different paradigms (or research programmes) may well find their explanatory frameworks incommensurable, but they would have to agree on these low-level established results as something needing explanation.

6
Science as a Social Activity

Scientific knowledge must be public and consensible.

6.1 Mary's Butterflies

Here, we meet one of Mary's descendants, also called Mary.

> Mary was sitting in her university office in front of the computer screen, the same way she spent most hours of most days. Sometimes she wondered if people outside of academia realised how much time scientists spend reading or writing, rather than doing experiments in the laboratory or making field observations. She often envied her great-grandmother of the same name, her great inspiration in life. Old Mary had also studied butterflies and their host plants, but had done so mostly in her own garden. Moreover, for her there was no pressure to publish her findings unless she wanted to, whereas Young Mary simply had no choice but to do so if she wanted a career in science.

> Today Mary was struggling with the reviewers' comments on her latest manuscript. For some years she had been developing a hypothesis regarding butterfly host plant choice, with possible applications to other insects and perhaps many parasites as well. It was generally thought in this field of research that butterflies do not actually make a real 'choice' of host plants, in the sense that they first check out what is available and then pick a plant among them to put one or more eggs on. Rather, they use vision and smell to find a potential host plant, land on it, check the taste and texture and then either lay eggs or reject the plant, after which they fly on to find another potential host. So, a sequential judgement of hosts rather than a choice among simultaneous alternatives. However, Mary had seen a behaviour in several butterfly species suggesting that something like a real choice may be going on: females often fly around and seem to inspect several plants for some time at a site before landing on one of them. Over the last few years she and her group of PhD students and post docs had been testing this hypothesis, using structured field observations as well as experiments in flight cages in the laboratory.

This time she had thought that they finally had strong results in favour of her hypothesis that could convince even her more sceptical colleagues, so she had sent a manuscript to one of the most respected journals in the field. Two of the reviewers seemed to agree and were quite positive, having only minor comments about phrasings and some issues about statistics that she could address quite easily, but the third was of the opinion that Mary's experimental results actually was not conclusive at all, because the situation in the laboratory was too unnatural and not representative of what butterfly females do in the field. Mary took a deep breath and then started to write up her responses to the reviewers' comments.

The next day she was finished, after having also discussed the responses with her group. In particular she had tried to explain how the results from her laboratory experiments were valid, given the close similarity to behaviour she had observed in the field, and how the more controlled situation in the cages actually added to the evidence for real host plant choice. Hopefully this would be enough to convince the reviewer and editor, so that the manuscript could finally be accepted for publication. She was planning to go to a conference in a couple of months to present her results and was looking forward to getting some feedback, but it would be nice to have the paper in press by then. Well, well, here goes nothing, Mary thought, and clicked the button to submit the revised manuscript.

6.2 Public Knowledge

The way that science is done and disseminated has changed a great deal in the time between the old and new Marys in our story in Section 6.1. The journey has moved on from the emphasis on facts, through hypothesis testing to the construction of increasingly stable theoretical structures, but as we have seen, none of these attempts to understand science succeeds in fully capturing what science is. Apparently, what is considered to be an appropriate scientific method is in a flux; disciplines can differ markedly in the details of methodology, and there simply is no universally accepted method of science across disciplines and across time. One possible conclusion at this point is that we should give up trying to define science at all. Perhaps there is nothing special about science? If you remember from Chapter 4, this was what prompted Paul Feyerabend to provocatively propose that the only thing we can conclude from the history of science is that anything goes. This critique of a universal method, from Feyerabend and other like-minded philosophers, has indeed been influential. Particularly in parts of the social sciences, where emphasis is on relativism and point of view, rather than adherence to a common method.

However, this isn't the only conclusion one can draw from the insight that there is no universally accepted scientific method across disciplines and across time. Let us pause and ask ourselves why we would expect that in the first place. We can see science as a continuously evolving set of insights, a machinery to generate knowledge about the world. As theories are adjusted and sometimes replaced, it is important as a scientist to stay on top of the theoretical development of their field. But not only that, a good scientist must also stay on top of the methodological development of the field. We have seen rather drastic changes over the years, not only in the nitty gritty details of how best to measure what we are interested in, but also in what is considered sound scientific practice. Using the terminology from Lakatos in Section 5.4.1, the standards of what is considered good science – the positive heuristics – differ between research programmes. This is true whether we make historical comparisons or if we compare contemporary programs. Sometimes these differences are small and subtle, sometimes they are more substantial. As Chapters 1–5 in this book demonstrate, even the very gold standard of proper scientific method has changed over the years. There has been progress, not only in scientific results, but in the way science is done. A static and universal scientific method appears to be a totally unrealistic and even undesirable goal that has been forced upon us by the needs of philosophers to nail down science, and distinguish it from other methods to gain knowledge. As scientists, surely, we want the scientific method to improve over time, just like we want our theories to improve?

6.2.1 The Problem of Demarcation

One of the issues you are faced with when trying to find a working definition of science, or even a general understanding of what science is, is the problem of demarcation. That is, how to make sure that the lines that can be drawn up against other methods of seeking knowledge are sharp and definitive, while at the same time not excluding phenomena that we would like to qualify as science. This has been a central problem at least since the time of Hume and Kant, and Popper considered it the most fundamental in the theory of knowledge. Presumably, this is also why philosophers of science have spent so much effort scrutinising historical instances of scientific progress. Feyerabend, for example, spent a considerable part of his book *Against Method* examining the Copernican Revolution, implying that if a notion of science cannot even make full sense of this important scientific breakthrough, of what use is it as a general description of science?

Being evolutionary biologists, we find this stance somewhat baffling. Just like current biological species, science did not drop down onto the world fully formed; it *evolved* out of pre-existing traditions. If we allow ourselves to draw out the analogy with biological species and speciation, there has long been a debate of how to properly define species. There exist several species concepts, but for each of them it is possible to find examples where it falls short. When teaching, we often point out to our students that this is a situation that we should embrace and welcome. Speciation, after all, is a *process*, and if it is, we should expect there to be phases where we cannot conclusively say if a collection of individuals makes up one species or two. They may be in *the process of becoming* new species – not yet completely reproductively isolated or not distinct enough to be regarded as two species but also not cohesive enough to be considered as one. Speciation is not an instant switch but a gradual process. Once this has been pointed out, it seems self-evident, but interestingly, it continues to be a stumbling block for students in evolution unless it is pointed out.

For the same self-evident reasons, we should expect a phase in the evolution of science where it was 'in the process of becoming'. The road that has led to modern science has been long and winding, and different aspects have been developed at different times. With this view, it becomes rather pointless to ask when exactly science arose as a human activity. It has been a work in progress over many centuries. It has roots in Greek philosophy as well as in European mediaeval theology and has borrowed aspects from other cultural spheres along the way. Eventually, sometime between the Renaissance and the Industrial Revolution, science became cohesive enough to be recognisable as such without much ambiguity, but to exactly pinpoint that time is as fruitless as pinpointing exactly when a species becomes two.

Just as it is self-evident that a complex cultural phenomenon such as science must have evolved, we should expect it to be still evolving. Scientific ideals, standards and methods are not the same now as they were around the Copernican Revolution or even 50 years ago. And they will probably be somewhat different 50 years from now. And this is fine. Science is a work in progress. To paraphrase Feyerabend, if a notion of science cannot even allow for the evolution of science, of what use is it as a general description of science?

It would appear then that if there is something that makes science special and that distinguishes it from other human activities, it has to be something else than just the 'scientific method'. John Ziman was a philosopher who was contemporary with Feyerabend and who also saw problems with defining science solely in terms of method. Perhaps less polemic by nature, this did not lead him to declare that anything goes or to devise an 'anarchistic theory of knowledge'. Quite the opposite, actually. Like Kuhn and Lakatos, Ziman shifted the perspective from the details in what individual scientists are doing to viewing science as a collective endeavour, but in addition, he emphasised the basic principles and societal structures that makes science possible in the first place [20, 21, 22].

6.2.2 The Growth of Science

People have always been curious; it is a defining characteristic of *Homo sapiens*. Inventions have been made that slowly but steadily increased the efficiency of whatever enterprise humans have entered into. Hunting, agriculture, construction, transportation – the list goes on. But the keyword is 'slowly'. Good ideas have spread, but from person to person, by demonstration or by convincing arguments. As a consequence, it took time for even the most brilliant of inventions to travel through the world.

According to Ziman, the rise of a more cohesive science in Europe after the Renaissance was not caused by a sudden scientific or philosophical breakthrough. To no small extent, the spark that set the stage for the evolution of what we today recognise as modern science came from another domain of society. It was Gutenberg's invention of moveable type and the printing press, and the continued refinement of the printing process that followed, that allowed the pre-existing scientific efforts to mature. This invention was a true technological revolution with huge ramifications across society, making it possible to rapidly print many copies of any text at a relatively low cost and to disseminate them to large numbers of people. Suddenly, conversations could be held also between distant scholars. Ideas could travel and spread; they could be built upon by others and (equally important) sometimes rebutted. In principle, this was not new – open dissemination of ideas among peers preceded mass printing – but the printing press dramatically changed its pace and the number of scholars who could partake in the discussion. Scientific conversations that used to be a matter of a few colleagues in closed settings could now involve whole communities of scholars across countries and continents, many of whom did not know each other personally. Scientific exchange became a public affair.

This is at the heart of Ziman's view of science. Science, he says, is *public knowledge*. But what does this mean? It is a pointed and simple statement that does need some clarification. It is not enough that scientific results are made public as such. Science is not just a public library. Equally important is the ongoing public discussion among peers about these results and indeed about how science should be done. A discussion that can be eloquent and mutually illuminating but also at times petty, jealous, unbalanced and maddening. Scientists, after all, are people. The important thing is that all people involved subscribe to some fundamental principles, where the most crucial is a strive for consensus. If you don't agree with me, I should try to convince you, and vice versa. And in the end, both of us need to be willing to bow to the better arguments of the other. For this to work, we also need to be in basic agreement over what constitutes a valid argument. This may also change, but at any given time, in any given field, we must agree on that for it to be possible to work towards a consensus among peers.

The public nature of science sets it apart from the related realm of technology and engineering. When we think of science, we often think of end products such as aeroplanes, mobile phones or solar cells. None of these would have existed without science, but they are all also results of long periods of targeted technological development. This development often takes place within companies and is typically guarded by secrecy. When a new type of technological solution is developed, its inventor tries to patent it to make sure that their achievement is not used by anyone else without compensation. This is in stark contrast to scientific developments, where making their findings publicly available is not just a way for scientists to be nice and let others use their results but is an integral part of how science works. To no small degree it is the very reason for the success of science. By making scientific findings public, they can be criticised, tested and built upon by other researchers much faster than if they were guarded like trade secrets. The end goal of science is not to build a product or to make money but to further our collective knowledge (which in turn makes technological development possible).

6.2.3 Scientific Institutions

As we followed Old Mary and her butterflies through this book, we saw how she learnt several scientific skills as she developed her scientific mindset. She learnt how to make good observations, to make hypotheses and how to test these with experiments. In the end she also learnt the importance of a scientific community where findings can be discussed and that a fruitful scientific discourse requires a common ground in order to not be lost in a constant debate over the fundamentals. Debate should be about the interesting challenges facing the field, not about the issues that have already been solved (or that are beyond our reach). When we met Young Mary in the beginning of this chapter, she envied her great grandmother and her less constrained approach to her research. But in order for science to grow out of its more haphazard beginnings, there was an increased need to coordinate the efforts of researchers, who could be spread across the globe, and – not the least – to educate new budding scientists like young Mary herself. If there are to be anything like paradigms or research programmes, their coherence must be maintained so that scientists can work towards solving the outstanding issues of the day, rather than having any scientific dispute that emerges leading to fragmentation.

It is easy to see that maintaining such a common ground could be challenging. How do you ensure that all participants in the discussion share the necessary knowledge about the topic at hand to fruitfully contribute? Or in other words, how is the cohesiveness of a paradigm maintained? As we have just learnt, the invention of the printing press gave Old Mary and her peers a way to disseminate new findings promptly to those concerned. Unfortunately, just as this technology could be used to spread knowledge, it could also be used to spread falsehoods. Consequently, as science grew, it became essential to have a system to weed out sloppy experiments and erroneous conclusions from the information that actually brought the scientific consensus forward.

Over the years, the scientific community has worked out a number of ways to handle this and related problems in order to enable a sound scientific discourse. We will discuss this in more detail in the upcoming chapters on science in practice, but a few words may be warranted here to illustrate the point. As universities solidified and multiplied, they served several important functions. One of these was to award academic degrees to those that had acquired the necessary skills of their field. In a sense, a PhD is a stamp of approval, an entry ticket into the scientific community. The university stakes its reputation upon the quality of the scientists they award with a PhD. Hence, the PhD is an assurance that the freshly baked scientist is properly trained and has acquired a scientific mindset.

The first scientific journals sprang out of various special-interest societies as a way to disperse findings to a wider audience. One of the first was the *Philosophical Transactions of the Royal Society* that was first published in 1665 (and is still being published!), and new journals were founded as fields proliferated. When the scientific community grew, discussions among peers during society meetings became inefficient as an error check, and eventually the peer review system evolved to fill the need of weeding out erroneous findings or unqualified speculation. Today all scientific journals rely on the critical scrutiny of other, typically anonymous, experts in the field to decide if the manuscript is a valid contribution that advances the current knowledge and if it is relevant for the scope of the journal.

Another consequence of the growth of science was that scientific work could no longer be reserved for people with large personal wealth. As a result, scientific work came increasingly to rely on external funding. Funds could come from private benefactors, but as scientific development was also increasingly seen as an endeavour of national importance, states around the world created funding agencies to disperse public money to promising researchers and projects. Again, these institutions typically also rely on independent peer review by experienced scientists of the applications as a means to ensure that funds go to projects that are scientifically sound and interesting enough to warrant funding. Thus, the scientific institutions evolved over time to facilitate the creation and spread of sound scientific knowledge. The solutions to the problems that arose, and keep arising, aren't necessarily perfect, but it is important that they work well enough to let scientific work proceed. The proper functioning of these institutions is so crucial that we can safely say that without them, there would be no science. It would not be too far-fetched to claim that in an important sense, science is its institutions.

While the scientific institutions serve several important purposes, the preceding paragraphs should have highlighted one function that is of particular significance. Science would simply not work if there were serious reasons to doubt the reliability of its output, and indeed, all scientific institutions play critical roles in maintaining what we call the web of trust. This web consists of several interrelated checks and constructs that all work together to protect the integrity of scientific findings. The importance of this cannot be overstated. As working scientists, we must be confident that published studies are performed according to current scientific standards. This is especially important when reading and citing papers from outside our own immediate field, where our own expertise may not be sufficient to make that judgement by ourselves. Since actors outside of the academy usually have even less ability to check the integrity of scientific results, trust in the process leading up to them is also essential for the development of working products and for shaping sound policy decisions. All scientists are active within this web, being a part of it themselves but also depending on its integrity for their work. The web of trust is what makes science work.

It may be worth pointing out that our trust in science is not – and should not be – blind. Scientists know very well that published results, or their interpretations, have to be viewed with a sceptical mind and that they will sometimes turn out to be erroneous. Still, with this in mind, we must trust that studies are performed and interpreted with an honest intention, and that they follow good scientific practice. We stated already in the beginning of the book that to a large extent, the fallibility of science is a strength. It is a built-in error correction that allows science to progress, rather than stagnate around established truths. Science encourages probing uncharted grounds, with the understanding that sometimes promising soil may turn out barren, but this error-checking would not work if there were no system in place to dissipate and scrutinise scientific results. It is also worth pointing out that the web of trust is a fragile construct. It relies heavily on the integrity of the institutions and of the scientists themselves. In fact, this is the main reason why scientific misconduct is considered such a serious offence. Whether it is an individual scientist fabricating data, or a scientific institution weakening its standards for financial gain, such actions threaten the integrity of science because they erode the web of trust.

6.3 A Social Definition of Science

It is time to introduce an alternative interpretation of what science is, one that is based on the insights outlined in this chapter. We have mentioned that John Ziman saw science as public knowledge, as opposed to private knowledge. Knowledge does not become part of science until it is published, criticised and defended. Ziman also stressed that science must be *consensible*, that is, for a question to be scientific, it must be possible for researchers to work towards an agreement on it. If it isn't, or if there is no striving for consensus on it, there can be no progress, since all individual researchers will only keep digging in their own rabbit holes.

6.3.1 The Search for Consensus

That science must be public and consensible seems like an important realisation, but is too vague to work as an actual definition. Nevertheless, if we want to capture what science *is*, what makes it stand out from other means of gaining knowledge of the world, it seems that these insights into the social and public nature of science need to be taken into account. A slightly more elaborate interpretation of Ziman's view of science that we have borrowed from Alan Cromer [23] is that science is *the search for a consensus of rational opinion among all competent researchers*. In spite of being a staggeringly simple definition (or perhaps because of it), it may need some further explanation to make sense. Let us take the three statements of the definition in order:

Search for a consensus. It is not the consensus itself that is the main goal (that would mean that the scientific pursuit has come to a halt), but the search for it. In other words, scientists should seek to influence other researchers but also be prepared to listen to their arguments. They should strive to reach a common understanding. If your peers disagree, you should try to persuade them. In practice today, publishing is the main way of attempting to affect this consensus. In a scientific article, the Introduction is where you outline the current consensus (corresponding to the current background knowledge), and what you consider lacking with it, and the discussion is where you argue for how this consensus should change as a consequence of your results.

Rational opinion. Not any opinion will do. To be part of science is to accept that if you want your arguments to be taken seriously, they have to be rational. This also means that you accept to give in to rational arguments of other people, should they be more convincing than yours. In practice, scientists can present rational arguments through results of experiments and other tests, and often by means of statistics. Importantly, what will be considered to constitute a valid rational opinion will vary somewhat from field to field, and from time to time. Note that this has the rather appealing implication that scientific fields that are not experimental can be included in this definition of science, as long as they provide other methods for assessing rational opinion, and as long as the theoretical framework for the scientific discourse is itself rationally based. It is thus not enough for an argument to be logically coherent, as you could form logically coherent arguments within an otherwise irrational framework.

All competent researchers. Who can participate in a scientific discussion? The statement 'all competent researchers' is both inclusive and exclusive in different ways. A competent researcher is someone who has demonstrated proficiency in the field.[1] In practice, today this is done through acquiring an academic degree, such as a PhD. Further demonstration of this competency is typically required – researchers often need to demonstrate their skills through a publication record and CV, and this record is an important aspect of evaluating the competence of a researcher. As we wrote in Section 6.2.3, a major purpose of university education, and of awarding academic degrees and titles, is to train and qualify competent researchers. It is possible for a person outside of academia to influence the consensus, but it is difficult, and would require some other means of establishing competency. The explicit use of 'all' is to emphasise that everyone that has acquired the necessary competence should be allowed to participate, and that arguments (such as scientific articles) should be made available for all

---

[1] Ziman himself (in *Public knowledge*, p. 9) used the phrase 'consensus of rational opinion over the widest possible field' rather than 'all competent researchers'. However, he also noted that only 'scientifically competent' persons should be allowed to influence the consensus (p. 63).

such researchers to scrutinise. That is, knowledge and the scientific debate, must be public.

We will return to these three statements, and how they relate to modern scientific practice, in Chapters 8–10 of this book.

6.3.2 The Social Definition and Philosophy of Science

The social definition just outlined approaches the problem from a rather different angle than most other approaches we have met. Yet, many of the attempts that have been suggested by philosophers to describe and understand science can actually be fitted into the three statements that form this definition. *All competent researchers* are those that are working within the same research programme or paradigm. What constitutes *rational opinion* will vary between paradigms/research programmes and will include the positive and negative heuristics of the current research programme. At different points in time and in different fields, the accepted methodology to rationally argue an opinion has to varying degrees included allusions to unbiased observation, induction, logic, falsification and experiment, among others. But importantly, this has changed over time and will probably continue to change. So, in comparison with the many attempts to define science in terms of method, the social definition is more future-proof.

The *search for consensus* is the truly social aspect of the definition, and it touches upon a crucial point: for an activity to be regarded as science, participating researchers must allow open discourse and aspire to a common understanding of any subject of controversy. The Bayesian approach delivered a formalised explanation for how researchers with diverging standpoints can move towards consensus in the face of new evidence. We noted in relation to Bayesianism that this was indeed an important feature of science. But it is not enough to provide a mechanism for reaching consensus. The *willingness* to work towards a consensus is arguably even more important. What if a field of research denies the search for consensus as an integral part of science, for example, by emphasising relativism and points of view rather than striving for objective knowledge? There are branches of some disciplines, most notably in the social sciences, that seem to have chosen this path. If they do, they deny themselves of a (rational) mechanism to affect each other's opinions. Their opinions would not be *consensible*, because if everything is a point of view, there is nothing to reach consensus on. Hence, they would not be compatible with the social definition of science. To the extent that this is an accurate description of these fields of research, they would then effectively have defined themselves out of this view of science.

On the other hand, the definition is agnostic with regard to method, and this opens the door for current and future fields of science that use different methods than the ones we have outlined in this book. One attractive feature of the social definition is thus that while it is actually pretty stringent with regard to what activities that will be accepted as science, it is flexible enough to allow scientific methods – indeed the basic standards of good science – to vary and to change. This is not to say that method is unimportant. On the contrary, it is very important for the success of science. Poor method makes for poor science, which is precisely why we need an active and ongoing discussion on how to apply the currently accepted scientific method but also to improve it. *How* science is done, however, is outside the scope of the social definition. It is a definition, not a prescription.

To illustrate the difference, let us return to the notion that any general description of science must be able to account for the Copernican Revolution. Feyerabend used this example as a wrecking ball to show how none of the methods proposed up until then passed this test of history. Researchers during the Copernican Revolution did not adhere to inductivism, positivism, falsificationism or any of the other candidates for a 'scientific method'. Earlier in this chapter, we questioned the view of science inherent in this statement, since it did not allow scientific methods and standards to evolve. The social definition is more universally applicable precisely because it does not specify methods or standards. We should not expect scientists during the Copernican Revolution to have followed the principles of falsificationism, for example, simply because falsification was an invention by Karl Popper in the twentieth century. It would only make sense if we think of Popper as a philosopher 'discovering' falsificationism, as something scientists have always been doing without realising it. But Popper's ideas came as a response to prevailing methods, mainly the then dominant logical positivism (itself an invention in response to previously prevailing traditions). Evidently, Popper meant that the scientific method of the time was lacking, and developed what has become known as falsificationism as an alternative. The inability to apply falsificationism to the Copernican Revolution is not a philosophical failure as Feyerabend would have it. On the contrary, it is an example of scientific methods and standards evolving to be better and more efficient at answering scientific questions. Still, science before as well as after Popper conformed to a notion of science as a search for a consensus of rational opinion among all competent researchers. He just challenged what 'rational opinion' should mean.

7
Synthesis

'It depends'.

7.1 What Do You Mean by 'Science'?

It seems we have to concede that the question of what science is may need more than one answer. It depends. We need different definitions of science because we actually mean rather different things when we say 'science'. The next time you are asked what science is, we suggest that you ask a follow-up question, because the questioner can actually seek the answer to at least three different questions that consequently require three types of explanations. They may want to understand what science is as a concept. What is the nature of science, and how is this different from other means of gaining knowledge? But they may also want a description of science, such as it is or has in fact been performed by scientists. Finally, they may want to know how science *should* be done. What is the best and most efficient way to learn about the world? We mean that all three questions are important to fully appreciate science, but they do need different types of answers. Perhaps one reason for why it has been so difficult to find a satisfactory answer to the question of what science is that it is not always clear which of these three questions we are actually trying to answer. As a consequence, explanations also tend to bleed into each other.

A straightforward way to make sense of these different explanations is to divide them into *conceptual*, *descriptive* or *prescriptive* explanations. Many of the approaches that we have met in this book have aspired to cover several of these types of explanations, but we would say that they typically emphasise one of them. The first examples that we met – the observation- and hypothesis-driven approaches such as empiricism and falsificationism – were mainly prescriptive. That is, they were mostly concerned with trying to ascertain how knowledge can be obtained and justified, and indeed what form scientific knowledge takes. In the process of doing this, they spelled out how scientific work ought to be carried out. They clearly also meant to provide a definition of science, although, as we have seen, such definitions are problematic since history has taught us that 'the scientific method' has always been a moving target.

The paradigm-driven approaches that we met next were mainly descriptive. They focused on describing the theoretical structures that scientists work within and how knowledge can fit into these structures, rather than on how knowledge should be obtained in itself. By doing so, they aimed to explain how these larger theoretical constructs influence and guide epistemological decisions (such as why researchers sometimes hold on to hypotheses in the face of apparent falsification). For better or worse, they claimed, we are all working within the boundaries of our current paradigm or research programme, which guides us but also constrains us to some extent. This was an important realisation that placed science, and scientists, in a social as well as historical context. Descriptions of science are useful and important, but they become vague and cumbersome if we want to boil them down to a definition. It would perhaps be possible to simply define science as an activity that *has* a paradigm, or a research programme, but since neither of these terms are particularly well defined themselves (Kuhn even argued that 'paradigm' cannot be precisely defined), such a definition will not be very helpful to truly distinguish science from non-science. Both Kuhn and Lakatos objected to their ideas being called 'merely' descriptive, but we would argue that this was their main contribution.

Finally, in Chapter 6, we met what we have chosen to call the social definition of science. We had found that it was problematic to define science in terms of method, because it makes it hard to cope with changes in method. For this reason, we argued that a definition of science needs to be method-agnostic and provided the social definition as an example of this. It is a conceptual definition, and as such, it can help us understand what science *is*, as a social activity, and what distinguishes it from other such activities. We believe that much of how science works in practice can be understood in the light of the social definition, something that we will expand on in the remaining chapters. Having said this, the definition is clearly also wanting in some respects, because even if one accepts that it does capture what science is, conceptually, it provides little operational guidance on how to actually do science, or on how to describe and understand scientific progress, both of which are arguably of fundamental importance if we want to understand science. Instead, the definition can provide a framework to help us understand the other types of explanations. The prescriptive explanations deal with what we should consider to be rational opinion, while the descriptive explanations also address who a competent researcher is and how consensus can be reached in spite of the many problems that we have described with regard to asserting the truth of observations. Hence, the three types of explanations are complementary.

7.2 Search for Consensus as a Line of Demarcation

We noted in Chapter 6 how all method-based definitions of science struggled with the problem of demarcation, that is, how to properly distinguish science from other means of gaining knowledge. While we noted that the social definition escapes many of the traps inherent in method-based definitions, we have not explicitly discussed how well it actually delimits science. Does it fare any better than the other approaches in this respect? Does it provide a reasonable demarcation between science and non-science?

All fields within the natural sciences should comfortably qualify under this definition, which we hope should be clear from the preceding chapters. We also believe that it does exclude pseudoscience like astrology, simply because there is no search for consensus of rational opinion. Truth is typically already given in the starting assumptions, and there is no ongoing pursuit to modify the consensus based on new insights. It is worth repeating that it is not the consensus itself that is the goal, but the active and ongoing pursuit of it. Scientific consensus is not static but has to be constantly challenged and adjusted as new findings emerge. Pseudoscience typically fails at this, as well as in the requirement of 'rational opinion'. When evidence is encountered that is troubling for a pseudoscience, it is typically simply ignored.

So far so good, but if there is a line to be drawn between science and non-science based on this definition, where exactly should it be drawn? What about the humanities and the social sciences? While this book is primarily targeting the budding natural scientist, the question often comes up in our classes and a comparison with these fields can be instructive, not least because the methods can be very different. As we mentioned briefly in the beginning of the book, the most striking philosophical difference between these major fields is the plurality of approaches that we can find in the humanities and social sciences, or in a cross-disciplinary field such as psychology (Figure 7.1). While there are differences among natural scientists in, for example, how much they emphasise theory or observations, for the most part the basic approach to science is still rather similar across the diverse fields that make up the natural sciences. This is not at all the case in the humanities or the social sciences, where you can often find drastic differences. A first distinction can be made between what is sometimes called *analytic* and *continental* philosophy, a somewhat

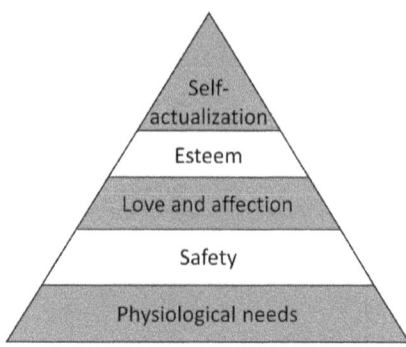

Figure 7.1 Example of a classic theory in psychology, a discipline which spans from the natural to the social sciences. This is the 'hierarchy of needs' postulated by Abraham Maslow [24], from physiological needs such as hunger (that must be met first) up to higher-order needs such as fulfilling one's goals in life. Be it right or wrong, it is a scientific theory because it is possible to search for a consensus of rational opinion regarding how well it explains human motivation for behaviours. Attempts have been made to test it empirically.

awkward classification that has its roots in the distinction between Kantian philosophy and British empiricism in the eighteenth century. Most of what we have discussed in this book would fall into (or follow from) analytic philosophy. Continental philosophy follows in the footsteps of the eighteenth-century German philosopher Immanuel Kant but has branched out into a complex set of philosophical approaches that includes phenomenology, existentialism, hermeneutics, structuralism, post-structuralism, deconstruction, psychoanalytic theory, postmodernism, critical theory, discourse analysis and so on. Most of these can also be understood as qualitative approaches, as opposed to the quantitative approaches that are more familiar to a natural scientist (and that are also used within some branches of the social sciences and humanities).

For a natural scientist, some of these research traditions may feel alien, but there are some valid reasons for using other approaches than in the natural sciences. The study of human expression, of human behaviour and of the institutions that make up our societies stands out because they all involve direct and indirect actions of conscious beings and interactions between them. Natural scientists rarely have to deal with intentions, interpretation and meaning, for example. Moreover, while the natural sciences almost always search for generalities (even 'laws'), many questions in the social sciences and humanities explicitly investigate unique phenomena. Scholars that do study such questions argue that the quantitative approaches like those used in the natural sciences are not suited to investigate these subject matters. Thus, different approaches may be appropriate for different questions. Still, the large variety of approaches could in itself make a search for consensus more difficult, since comparing insights becomes more challenging.

We do not want to dive too deeply into these approaches, but a quick overview may shed some light on how the 'social definition' would apply to subjects outside of the natural sciences. To this end, we have chosen three prominent approaches to exemplify. We want to emphasise that this is not intended as a critique of these approaches and their usefulness in their respective fields, only as an exercise in demarcation, given the definition at hand. The first example is *phenomenology*, which traces a rather clear ancestry to Kant's distinction between the thing-in-itself and how such an object manifests itself to the human observer (the thing-for-us or the phenomenon). Phenomenology was developed as a critique of the prevailing positivist tradition and was pioneered by the German mathematician and philosopher Edmund Husserl around 1900. Husserl meant that while we do not have access to the external reality, we can understand it by describing how it manifests itself to us (recall our observation of an oak tree in Chapter 1). Such a manifestation – the phenomenon – is not the thing-itself, but importantly also not simply a subjective sensation. The phenomenon emerges from the interaction between the object and the observer, and the term intersubjectivity is commonly used to describe this. It is perhaps most easily understood when the 'object' of study is another person, but it can also help bridge the gap between the subjective minds of different observers. To give just one example, Kemkes and Akerman [25] performed a phenomenological study in the Chequamegon Bay area in Wisconsin, USA, interviewing 17 residents in the area in depth regarding how they felt about climate change. Among the results highlighted was that the failure of global collective action produces feelings of helplessness and anxiety and that public forums for discussion are needed.

By careful descriptions of how a phenomenon manifests itself in all its different aspects, both to yourself and to other observers, phenomenologists argue that it is possible to achieve an objective understanding of its essence. The goal here is clearly different from the natural sciences in that the ambition is not so much to reveal causality or even generalities but an accurate and meaningful description, as well as a deeper understanding of the issue under study. Still, insofar as there is a clear strive for consensus (there are several different branches also on the phenomenological tree, and not all seem alike in this respect) and a set of explicit and implicit rules as to what constitutes rational opinion, phenomenology would pass the line of demarcation according to the social definition of science.

The second example is *hermeneutics*, which has its roots in biblical exegesis, that is, the critical interpretation of religious texts. Modern hermeneutics has developed to include the interpretation of any kind of text, as well as non-textual and non-verbal communication. Hermeneutics makes a distinction between what is written and what is meant by it, and the preconceptions of the reader are important to consider. Since prejudices will always be part of the interpretation, they need to be made explicit and are indeed integral to the method. By, for example, reading a text several times, the preconceptions will change, and new meanings may emerge. This dialogue between text and context is called the hermeneutic circle and will enhance and elaborate the interpretation as the reader proceeds through it. Again, hermeneutics is itself a tree with many branches, but at least some of these branches seem to reject relativism. The conviction that our interpretation of the world is mediated by our preconceptions does not mean that we are making the world up. It appears to us that hermeneutics falls into a grey zone with regard to the social definition. There is a strong emphasis on subjective readings, but it does allow for a rational search for consensus of the interpretations of these readings.

The final example is the *postmodernist* movement. This is a more recent development, arising in the twentieth century as a criticism of the modernist movement within literature, art and philosophy that preceded it. Postmodernism (like modernism) is a sprawling set of ideas that is difficult to fully grasp. It is characterised by a rejection of notions such as universal theory and even the concepts of truth and reality. Instead, postmodernists embrace relativism and individual narratives, contending that claims of truth always hinge on point of view and the prevailing power structures. You may recognise some of this from Paul Feyerabend, whom we have mentioned a few times in the book, and he is indeed often placed within the postmodernist philosophical tradition. In our view, it is difficult to reconcile this approach with the social definition of science. If you reject the existence of an objective reality and instead focus on points of view, the search for consensus almost by definition becomes a meaningless pursuit, and further fragmentation seems inevitable.

We will stop there, as we think this excursion outside of the natural sciences is enough to illustrate that many of the research traditions in the social sciences and humanities clearly have the potential for a formalised search for consensus, just like the natural sciences, even if they use very different methods. Since the social definition is method-agnostic, it is open to alternative ways of arguing a rational opinion. As a definition of science, we believe this is a distinct advantage over alternatives that are more or less rooted in the methodological approaches of the natural sciences (or any other methodological approach for that matter). It is interesting to note that to the extent that a discipline does not pass the bar, it is often by design. The reason for why a formalised search for consensus is not always realised seems to be that there can be disagreement on whether this is actually a desirable goal. It should also be evident from this brief sketch that to the extent that the social definition provides a line of demarcation, it is not particularly sharp. There is a grey zone, but it may well be that the borders of science *are* fuzzy.

7.3 Being Scientific

It would seem then that the idea of a line of demarcation between science and non-science is not straightforward. Indeed, if we are correct in concluding that the concept of science has evolved over historical time, it follows from this conclusion that the line of demarcation cannot be very sharp. However, there is another dimension to this discussion that must be recognised. Even if there is no clear-cut division between science and non-science, some human activities purporting to find and disseminate knowledge can clearly still be said to be more 'scientific' than others. At the very least there is a spectrum (Figure 7.2) from activities that everybody would call science, over those where it can be hard to tell at first glance, to those activities that most people would see as distinctly different from science. The intermediate activities range from *quasi-science*, activities with some scientific aspects, such as a car mechanic experimenting to figure out why the car won't start, to *pseudoscience*, which tries to emulate the superficial aspects of science in order to gain credibility. We would argue that religion and mythology lie at the opposite end of the spectrum from science, but more on this in Section 7.3.3.

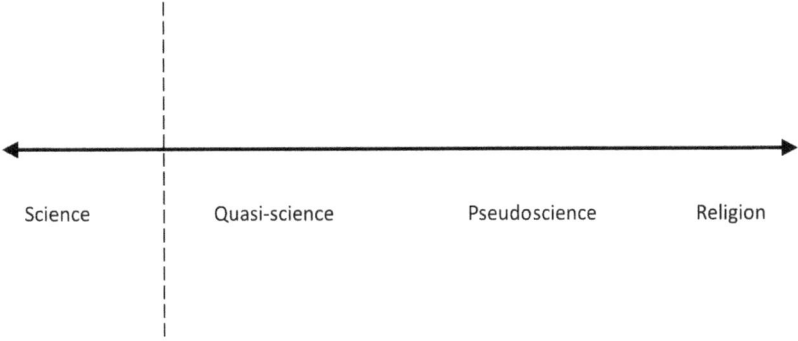

Figure 7.2 Some different human activities with the ambition to produce and/or give out knowledge about the world, placed on a spectrum from the most scientific to the least. Here 'quasi-science' refers to activities with some scientific aspects but not fully qualifying as science, and 'pseudoscience' refers to non-scientific activities misleadingly presented as science. Where exactly is the 'line of demarcation' (dashed line) between science and non-science? Is science simply what scientists do?

Moreover, even science activities can conform more or less to a scientific ideal. This ideal extends beyond the scientific method and is more related to what it means to be a 'competent researcher'. In Chapter 6, we emphasised the importance of the academic institutions for maintaining the integrity of science and its web of trust. In line with that, one could argue that if a clear line of demarcation exists at all for modern science, it should be tied to being part of this academic world and its scientific publications. This is what we meant in Chapter 6 when we wrote that, in an important sense, science is its institutions. Most would probably agree that a professor with a PhD doing and publishing research within the academic system is a scientist, and thus the research that they do and the published results are 'by definition' science, but their precise way of doing it can nevertheless be more or less scientific and thus more or less 'good' science. So how do we recognise good science? What does it mean to be 'scientific'?

7.3.1 The Marks of Good Science

Some of the most often acknowledged marks of good science are repeatability, transparency, objectivity, parsimony, willingness to test ideas, willingness to abandon ideas and a non-authoritarian, equal and open approach to knowledge.

- **Repeatability.** As a sign of good science repeatability refers to the reasonable notion that if observations and experiments can be repeated with the same or at least similar results, they should be more reliable as indicators of how the world is really like. If it is to be possible to reach consensus on the veracity of an observation, it must be possible for others to verify it by repeating it. As we saw in Chapter 2, it is an important part of empiricism that observations can be verified this way. The hypothesis testing approach introduced in Chapter 3 is less adamant about repeating observations and experiments, in particular when it comes to early Popperian falsificationism with its emphasis on how a single observation can falsify a hypothesis. However, we have argued that this may be easier said than done, and in practice repeated 'falsification' (often in different ways) is typically necessary before a hypothesis is completely abandoned. Moreover, Popper stressed in his later writings that results must be repeatable *in principle*, even if they are never actually repeated. In other words, the methods used to produce the results must be documented and reported in such detail that another researcher could evaluate the methods

and conclusions and, if necessary, repeat the observation to see if they obtain the same result.
- **Transparency.** In order for research to be repeatable, scientific work must be transparent. But transparency is important for other reasons as well. If data, procedures and the statistical code behind tests are made publicly available along with the publication, it is possible to not only go back and verify the results but also to reuse the data and build upon it in innovative ways. Transparency with funding, affiliations and other potential conflicts of interest also aids in evaluating how trustworthy the conclusions are and how independently they were reached.
- **Objectivity.** Scientists should strive to find out the truth about an objectively existing world and thus try to be objective in their research, eliminating subjectivity as much as possible – for instance, by performing controlled and well-designed experiments. Every attempt should be made to counter any bias of interpretation resulting from expecting or hoping for a certain result, if possible by blinding experimental treatments until the results are ready.
- **Parsimony.** The concept of parsimony was introduced in Chapter 2; a good scientist should not postulate a complicated hypothesis if there is a simpler alternative (one making fewer assumptions) that explains observations equally well.
- **The Willingness to Test Hypotheses.** This is crucial in science. It is an integral part of the hypothesis testing approach of Chapters 3 and 4, but also empiricists like Francis Bacon in Chapter 2 would agree that observational results must be tested by repeating them under varying circumstances to see if they really are general. Moreover, it is at the core of science as a social activity (Chapters 5 and 6) that proposed hypotheses will also be tested by other scientists once the results have been published.
- **The Willingness to Abandon a Hypothesis.** A good scientist must also be willing to abandon a hypothesis if it is not corroborated, at least when enough evidence or theoretical arguments have accumulated that suggest that it is not correct. We mentioned already at the very beginning of the book that the acknowledgement that 'I could be wrong' is something of a cardinal rule in science.
- **A Non-authoritarian Approach to Knowledge.** Finally, the scientific ideal suggests that it should not matter *who* presents scientific results, as long as the science is sound. Results presented by authorities, such as established professors, should be scrutinised as strongly as those from a

new PhD. Furthermore, there should be a free and open discussion of ideas so that the currently best-supported ideas can have the most impact on future research.

7.3.2 When Science Is Not Scientific

Perhaps needless to say, science and scientists do not always conform perfectly to these ideals. There is always a temptation to publish results too early, perhaps because they fit expectations from the scientist's pet hypothesis, and it may turn out that the results were in fact a fluke and cannot be repeated. Subjectivity may have influenced these results as well as the decision to submit them for publication. Another possibility is that the results were in fact not exactly as expected, but the pet hypothesis was 'saved' by making it a bit more complicated and less parsimonious, adding assumptions. As discussed in Section 3.4, such post hoc modifications can be defended if they are independently testable, but this is not always the case. If the added assumptions are more of an ad hoc character ('some instrument readings did not come out as predicted, but this may have been because of power fluctuations') they compromise both the actual willingness to test hypotheses and the willingness to abandon them. Authority can still matter in science, but it can (and should be) discussed to what extent it is reasonable that results presented by an established scientist could be intrinsically more reliable than those from a newly baked PhD. A greater problem is arguably when perceived authority stems baselessly from, for example, gender or ethnicity. Finally, scientific discussion is not always equally open to all researchers due to, for example, language barriers and access to costly library resources.

7.3.3 The Marks of 'Non-Science'

It is an interesting exercise to reverse the criteria for good science given in Section 7.3.1, as this illuminates the other end of the spectrum proposed above (Figure 7.2). Pseudoscience typically has no problem with poor transparency and lack of repeatability of results. If there are anomalies, they are explained ad hoc. In, for instance, fringe branches of 'alternative medicine' it is often said that all situations are unique, preventing the use of normal experimental methods with a controlled design. To take it to the extreme, the basis for religions are unique phenomena and events: gods, miracles and other divine interventions. The lack of objectivity in pseudoscience is shown by the use of selected examples supporting the central claims, even when it would be possible to make more controlled observations or even experiments. There is a general unwillingness to test ideas against reality, and when facts and results counter to the theories are met upon or put forward by opponents, they are downplayed or ignored. If alternative theories are proposed, these are dismissed even when they are better explanations of observed phenomena. There is also typically a strong belief in authorities, they are the ones who know what is true or not, and there is typically little transparency in how these authorities arrived at their opinions. Again, religion seems to fit these descriptions even better, with its basis in divine authority and its emphasis on faith over evidence, even (or in particular) in situations when the divinity could be questioned based on observations or lack of observations.[1] This is not intended as a value judgement on religion, only as a suggestion that religion is very different from science as a way of obtaining knowledge about the world.

7.4 The Web of Trust

Although it would be tempting to see the marks of good science given in Section 7.3.1 as an alternative way of defining science, it is clear that *science* and *scientific* do not always go hand in hand. As already mentioned, it is perfectly possible to have a scientific approach to work outside of academia, such as car repairs or cooking, and for scientists to have an, in some respects, unscientific approach to science. Adherence to the marks of good science is an important goal for all scientists, but as we hope this and the preceding chapters have made clear, *science* is something more than this. We wrote in Chapter 6 that science is a machinery for generating knowledge, and a good scientific attitude is but one important cog in this machinery. For science to work efficiently as a generator of new knowledge, a number of other supportive cogs must be in place as well and work towards the same goal. This supportive system is what we throughout the book have called the web of trust.

The web of trust should not be taken for granted. It can easily be eroded, and the supporting institutions need to be designed around the goal of maintaining and justifying this trust. It is also important for the individual scientist to not lose sight of how essential, and how fragile, the web of trust is. In the highly competitive and cut-throat environment that science often is these days, a high-ranked publication can make or break a scientific career. Hence, the temptation to 'fix' data, or otherwise enter an ethical grey zone, has probably never been

---

[1] See, for example, John 20:29: 'Because you have seen me, you have believed; blessed are those who have not seen and yet have believed'.

higher. It is therefore imperative to understand that venturing into this grey zone threatens the very fabric of what makes science work.

However, it should also be clear by now that the web of trust is complex and not always easy to comprehend. We must learn what we can trust and why some sources are more reliable than others. This is true for budding scientists in academia, but to a large extent also for anyone who needs to evaluate claims made to 'follow the science'. Which today basically means everyone. As a consequence, it has now become time to take a closer look at how the web of trust works in practice in contemporary science – keeping in mind that it is a fluid system that may well change in the future. For example, the rapid development of AI-based tools may turn out to be disruptive for many aspects of the web of trust, from the generation and analysis of data to the publication system. These potential changes are still unknown and difficult to predict, and in any case, a snapshot of how the web of trust currently works should be illustrative. As we have argued that the 'social definition' provides a good framework to understand how science works, we will cover its three main components in turn. 'A search for consensus' requires a public discourse, and Chapter 8 will focus on how scientific knowledge is made public through the system of scientific publishing. In most natural sciences, the generation and analysis of data are important aspects of what it means to argue a 'rational opinion'. Hence, in Chapter 9, we cover the nature of scientific data as well as the use of statistics and show how statistical choices ultimately are philosophical decisions. Finally, in Chapter 10, we discuss what it means to be a 'competent researcher', and argue that the competent researcher has an important role to play also outside the academy.

8
Science in Practice: Publishing

The search for consensus: making knowledge public.

8.1 The Publication Process

In modern science, publishing in scientific (peer-reviewed) publications is the primary way of attempting to influence the prevailing 'consensus of rational opinion' in the field by adding new knowledge and scrutinising previous findings. The peer-review process is an important part of the 'constructive scepticism' that is an integral part of science, combining the two pillars of science once pointed out by Jacob Bronowski: trust and dissent [26]. Hence, peer-reviewed publications are also the most important academic merit for obtaining positions and research grants. Other forms of publications, such as popular science or reports ordered by governmental agencies are not given the same weight as merits, because of the lack of scrutiny by peers. Talks at conferences and invited seminars are likewise of importance both to disseminate results and as merits, but reach a smaller audience and are not scrutinised to the same degree. The central importance of scientific publications in academia warrants a closer look at this phenomenon to see if scientific theory can shine a light on some of its aspects.

It should be noted that what we will describe here is how the publication process has worked for several decades, but this does not mean that it will not change in the future, as it has in the past. In fact, this is one of the aspects that is perhaps most likely to change as a consequence of the rise of AI, but also the increasing discontent with for-profit scientific journals exploiting the free work of scientists as authors, editors and peer reviewers. Nevertheless, we believe it is important to understand the various functions of the publication process as it exists today, because whatever will replace it must in some way also replace these functions if the web of trust is to be held intact and science continue to be a reliable source of knowledge.

8.1.1 The Format of Scientific Publications

The French *Journal des scavans* ('journal of the learned') and the English *Philosophical Transactions of the Royal Society* were both first founded in 1665 and are generally recognised as the first scientific journals. However, peer review as it exists today came later, and if we use this common criterion for what counts as a 'Scientific Journal', the phenomenon itself emerged over time, as did the precise format of scientific publications commonly seen today. In fact, it was not until the late twentieth century that peer review came to be seen as a requirement for scientific legitimacy. It has been suggested that this happened in response to attacks on scientific funding in the Cold War United States [27]. The peer review process was publicly elevated in order to reinforce and spread the belief that only scientific experts can decide what is good science. This well illustrates the conclusion that we arrived at earlier in the book, that the definition of 'science' has evolved over time and is not historically static. Moreover, in some disciplines (notably physics) non-peer-reviewed pre-prints in digital archives (more on this format in Section 8.1.3) do have a status as scientific publications, illustrating that the concept of science is also variable at a given time in history.

There are some variations to the modern format of science papers, but very often it follows the structure Title, Abstract, Introduction, Materials and Methods, Results, Discussion, Acknowledgements and finally a list of References. These sections of the research paper can profitably be seen as having different 'functions' as they relate to the academic discussion in the field of research:

8.1.1.1 Title

The title has the function of reflecting the content of the paper in a very condensed manner, in such a way that it is possible to immediately judge whether the content is of interest to another scientist. Every active scientist is flooded by new publications and does not have time to read more than a small fraction even of those within their particular field. Thus, it is imperative that it is possible to find those papers that should be scrutinised further in order to stay up-to-date with emerging knowledge in the field, and to find them just by skimming titles. It should also be possible to easily find older publications when the need arises, from a list of titles stemming from a search in a literature database (see Section 8.1.4).

A secondary function of the title is to attract interest and citations from other researchers, for instance by including current 'buzz words' in the field or by attempting to convey the novelty, conclusiveness and general importance of the results, already in the title. In this there is a natural temptation to overstate the results, and at least partly for this reason one of the tasks of the peer reviewer is to check that the title actually correctly reflects the content of the paper and is not misleading.

Titles can be either merely indicative of the contents ('Origin of basalt magmas – an experimental study of natural and synthetic rock systems') or actually present the main finding ('Intestinal mucin is a chaperone of multivalent copper'). The latter is more common in some fields than others and may perhaps reflect something of an empiricist tradition (Chapter 2), where scientific results are viewed as provisional facts – rather than as theories that have now gained some support. Indicative titles instead often signal a hypothesis testing, more theoretically exploratory tradition (Chapters 3–4).

8.1.1.2 Abstract

The abstract (or summary) has a similar function to the title, in that it should reflect the content of the paper in an abbreviated manner, so that once the title has caught the interest of the reader, the abstract can next be used to see whether the paper is actually important to read in more detail. In addition, the abstract itself should be a mini-version of the paper and thus reflect the motivation for the study, as well as a hint about the methods, results and conclusions, meaning that in many circumstances it may be enough to read the abstract to be aware of the publication and its 'take-home message', saving the actual reading for when there is time and necessity.

8.1.1.3 Introduction

This section has the important function of setting the scene for the paper as a contribution to scientific discussion in the field and typically follows a rather strict structure. It starts broadly with a sentence or two defining the research topic and why this subject is of general importance, central to the field of research, novel, interesting, topical or any combination of these claims. The formulations vary depending on the intended journal; a high-impact journal with a broad scientific audience may demand bolder claims than a specialised journal with a specialised audience, as the latter will be at least somewhat interested in most papers in the journal that are good enough to merit publication.

Next follows a paragraph or two outlining what is currently known about the subject, in other words, the 'background knowledge' of Chapter 4. This takes the form of a mini-review, with many citations of earlier publications. It is among the duties of the peer reviewer to check whether the most relevant publications are in fact cited, so that the new study has been done against the accepted background knowledge and not in reference to the authors' own (perhaps too subjective and/or incomplete) version of the state of knowledge in the field.

Following this outline of what is already known comes a crucial statement about what is *not* known, sometimes referred to as the 'research gap' or 'niche'. Such a statement is obligatory because it is the motivation for the study. If it is not possible to say what it is we don't know, there is no reason to perform the investigation in the first place. In more empiricist research traditions (Chapter 2) such statements can simply be about the need to somehow extend present knowledge ('this has been frequently studied in the Atlantic Ocean but never in the Pacific'), whereas in hypothesis-driven research (Chapters 3–4) they are more specific, leading up to the hypothesis or hypotheses that will be tested in an attempt to fill the research gap, and a hint on how this test will be performed.

In some fields, the Introduction ends with a sentence or two stating the main result(s); in others, this is left for later in the paper.

8.1.1.4 Materials and Methods

This section describes precisely how the scientific study was performed and has the dual function of making it repeatable for other researchers (at least in principle) and ensuring that it is possible to evaluate whether it was done using methods that are actually appropriate for making the claimed observations or for testing the hypothesis or hypotheses under investigation. For the latter reason, the peer reviewer will thus read this section very carefully.

Methods sections can be very intricate and detailed when the methods are new and designed for the purpose of the current study or somewhat more superficial when only standard methods are used, simply citing the original publications describing these methods.

8.1.1.5 Results

The Results section is the core of the paper, with the obvious function of presenting the results, doing so in text, figures and tables. In many fields, results are statistically investigated in order to check the statistical significance of results given variation among samples (Chapter 9). In more empiricist research traditions, results are often uncommented; they just are, whereas in more hypothesis testing traditions comments can indicate whether a particular result supports the hypothesis under investigation or was unexpected. In some journals, Results are even interwoven with the Discussion, going back and forth between the two, perhaps eliminating different sources of error or alternative explanations one at a time, testing a sequence of hypotheses, or seeking additional support for a particular hypothesis.

8.1.1.6 Discussion

This section of the paper is where the results are placed back into the context outlined in the Introduction, with the function of discussing how the results change the current 'consensus of rational opinion' (Chapter 6) or instead lend further support to some aspect of 'background knowledge'. The Discussion most often opens with either presenting the main results or by first reminding the reader of the 'research gap' from the Introduction, that is, the goal of the paper. The structure of the rest of the Discussion is variable and governed by how clear and complex the results are. If a single hypothesis was tested, it is enough to remind the reader of the hypothesis and then discuss if (or to what extent) the results support the hypothesis or not. If there were several hypotheses, they are often discussed in turn. Papers in the empiricist tradition instead discuss each main result in turn.

Either way, the Discussion often cycles back and forth from a specific result to its more general consequences and what further research may be needed, and then onto the next result. For each result any limitations of the study preventing firm conclusions can also be raised by the authors in order to demonstrate awareness of the limitations and thus not the least forestall criticism from peer reviewers ('Despite the low sample size, it is possible to conclude that ...'). The authors can also make an argument about why the limitations are not as severe as they might seem, grounded in the knowledge that they have about the study situation but that reviewers and readers may lack.

Some speculation about the general consequences of results, not strongly grounded in the results themselves, is permitted in the Discussion. This should, however, be clearly signalled to the reader: 'It is tempting to speculate that ...'.

An interesting shift in tense can often be seen at one or several points in the Discussion, from past to present tense, reflecting the theory-dependence and uncertainty of observations discussed earlier in the book. The outcomes of the study are first presented in the past tense in a very specific manner, directly relating to the study: 'Genes relating to stress were upregulated at the two highest experimental temperatures'. However, in order to move forward to more general consequences, the results must be provisionally accepted as real, even as approaching 'facts'. This is often done via a bridge signalling the move but also the remaining uncertainty: 'If this is a general pattern ...'. Once this move has been made, present tense can be used: 'The upregulation of stress genes indicates that ...'

The Discussion usually ends with a Conclusion section where the 'take-home message' is put forward. In most fields, this is the most general part of the paper, mapping out how the results may change current thinking in the field – linking back to the likewise general beginning of the Introduction. In some fields of research, such as medicine, the Conclusion is instead often very narrow: 'We conclude that drug X is more effective in alleviating malaria symptoms than drug Y'. The difference relates to whether the field is more theoretical and exploratory in nature or aims to produce immediately useful knowledge.

8.1.1.7 Acknowledgements

This is a short section with the function of giving the authors an opportunity to thank persons who may have contributed to the study, but not in a manner that warrants co-authorship. Funding agencies and infrastructures crucial for the work are also typically acknowledged here.

8.1.1.8 Author Contributions

Many journals now include a section to list what the individual authors have contributed to the paper. This is prompted by an ongoing trend in many scientific fields to have more multi-authored papers. Some articles can have *very* long lists of contributing authors, but it is quite common to have 5–10 authors. Requiring these to explicitly state their contribution increases transparency, but since authorship is such an important currency in the academic merit system, it is also an attempt to reduce the temptation to include 'authors' who had very little to do with the production of the paper.

8.1.1.9 References

Here, all cited references in the paper are listed. Citations have three main functions. First, in academic writing, all statements must be backed up, either by the data produced in the study or by citing earlier publications establishing the background knowledge. Second, the references provide a way for the reviewers and other readers to assess how strong such existing evidence is, as well as finding publications that they may have been unaware of. Third, citations are a way to give credit to earlier researchers in the field, acknowledging their contributions to shaping the current consensus.

8.1.2 How to Read Papers Effectively

The preceding section hints at some tips for reading a research paper effectively, in the sense of quickly absorbing the information it may contain. The Title and Abstract are obviously of central importance and should always be the first stop. These sections are a good indicator of whether the paper is likely to contain the information you are looking for, and if you are just exploring a topic, the information in the Abstract is sometimes all you need.

If you need more information about what was actually studied and why, look for the part of the Introduction where the 'research gap' is pointed out and offered to be filled. Very often you find a sentence starting with something like 'In this paper ...' or 'Here, we ...', that is, referring directly to the manuscript. This signals the intent to attempt to fill the knowledge gap and is followed by a specific description of how this was done. Just before this sentence you should find a clear description of the gap itself. However, sometimes the gap is instead only implied, embedded in the offer to fill it: 'In this paper we develop a new theory to explain the relationship between income and happiness' (suggesting that this is not yet well understood).

If, after having read the Abstract, you are still uncertain about the conclusions of the paper (an Abstract can be mostly indicative of the contents, especially if the conclusions are not straightforward), you could jump ahead to the end of Discussion to see if there is a Conclusion section, or a paragraph or two to that effect but without a subheading. The end of the Introduction and start of the Discussion sometimes also contain conclusions about the main results.

There are, of course, situations when the strategy just described is not enough, because you need more detail. This is particularly true if the research area is for some reason controversial, and then especially when the results are of direct importance for the general public (for instance, public health issues). In those situations, your strategy depends on what you need to know. Details about the methods will be found in the Materials and Methods section, and detailed results will be found in the Results section. These sections can be skimmed in a search for the information you are looking for. If you are interested in a particular result that caught your interest, you can next look for the part of the Discussion where this result is mentioned to see how it is interpreted by the authors. You might also want to go to the corresponding part of Materials and Methods to understand better exactly how the result was arrived at.

8.1.3 Publishing

The scientific publishing process itself is quite unique and opaque to the non-scientist, but many of its aspects can be understood from the notion of science as a 'search for a consensus of rational opinion among competent researchers'. In particular, peer reviewers can be seen as having the function of acting as academic 'gate-keepers', helping to ensure that published results come from 'competent researchers' and follow a rational basis for how they arrive at and interpret results, thus being worthy of publicly attempting to influence the present consensus. Without such scrutiny (or an equivalent mechanism), we could easily become flooded with misleading results of poor quality, impeding scientific progress. Here we outline the different steps in the publication process.

8.1.3.1 Pre-prints/archives

It has become increasingly popular to archive manuscripts in a public depository before peer review and publication in a scientific journal, so-called pre-prints. In some disciplines, such as physics, this is sometimes more important than publishing in a journal. Here, scientific scrutiny is performed by peers commenting on the archived version rather than through traditional peer review. Contributions deemed correct and of some importance by peers can be cited with reference to the archive, whereas flawed or non-significant contributions are ignored or commented on to this effect. An advantage of this is that more than two or three peers can look at and comment on the paper, but it requires an active community, and since no formal decision is taken on whether the manuscript should be 'accepted', it can be harder to evaluate which papers to trust. However, in most disciplines, the goal is to publish (hence the name '*pre*-print') and the archive is a step on the way, where it is possible to cite the manuscript and also to get comments that may improve it. Note that pre-prints have not been peer-reviewed and thus have a higher risk of being found faulty than publications in journals. They should thus be used with great caution as a source for knowledge, especially by someone not in the same field of study.

8.1.3.2 Choosing a Journal

There are tens of thousands of scientific journals, and even within a given discipline, there can be many to choose from. How do you make this choice? This is one of the many aspects of science where experience is important, experience which is transferred from supervisors and other more senior colleagues to PhD students as they learn to become scientists themselves.

Most importantly, a difficult judgement has to be made regarding whether the content of the manuscript is likely to interest a broad scientific audience, a more specialised audience or something in between. This decides what type of journal to send it to, as well as having a strong influence on how to format the manuscript for the intended audience. Reporting the discovery of the oldest fossil land-living vertebrate yet found – with a description of the fossil and the likely appearance of the organism – could go into a journal such as *Nature* or *Science* with their very broad scientific audiences and perhaps also attract media attention. If you or another researcher five years later re-interpret a part of the published description, perhaps showing that the suggested homology of some of the bones should be corrected (cf. Figure 5.1), this is mostly of interest to palaeontologists working on similar species. The same applies to a later discovery of another better-preserved fossil of the same species, equally old but much less interesting, illustrating how well-supported findings that clash with current background knowledge are rightfully viewed as of higher scientific significance (Chapter 4).

It should be noted here that publishing in more general journals gives more scientific prestige and merits. It is thus tempting to 'aim high', but where experience comes in is in making a good judgement of the manuscript so that not too much time and effort is wasted. First, the manuscript has to be precisely formatted following the journal's often detailed 'Instructions to authors' before it can be submitted, a not trivial task. Second, there is a cost in time and perhaps some damage to self-respect in having a manuscript submitted only to be rejected, not because there is anything wrong with it but because it is deemed not interesting enough for the journal's broad audience.

8.1.3.3 Types of Journals

In addition to the general-specialised continuum of journals, there are other matters to consider when considering a journal for your manuscript. Is it 'Open Access' (OA), that is, available online for everybody to read cost-free? Traditional journals are often published by scientific societies, and increasingly by commercial publishers, with individual scientists and university libraries subscribing to them for access to printed and/or online versions. There has been a strong movement towards publishing OA in recent years, and funding agencies and universities can demand that scientists publish open access or make the manuscript freely available by other means, such as placing it in a public archive. This came as a reaction to the often very steep subscription costs of traditional journals, meaning that scientific results were not available to the public (who may have funded the research via taxes) or to scientists in, for example, developing countries. If scientific results are hidden behind steep paywalls, it is hard to argue that it is public knowledge.

Freely available scientific knowledge is a positive development, helping to enable the free and open discussion of ideas that is a mark of good science (Chapter 7). However, the development is not problem-free. OA journals take their income from publishing fees instead of from subscriptions, meaning that it can be quite costly to publish, costs that have to be covered from research grants or universities. A layman may believe that scientists get paid for their publications, considering that they do all the work with writing and illustrations and most of the work with formatting the manuscript. This has, however, always been far from the truth, and even more so when it comes to OA journals.

This situation has led to the emergence of so-called predatory journals that live off publication fees but have little or no actual peer review, in effect publishing the work of anybody who can pay for it. Such journals take advantage of the need to publish in order to have an academic career and of the ignorance of scientists outside of established academic environments who may not realise that such publications have no real merit value.

With some experience, clear predatory journals are easily recognised, but there is also a grey zone of more scientific OA journals with peer review that have emerged to profit from publication fees. Some have the stated ambition to publish everything that is scientifically sound, whether or not it is 'interesting', and peer reviewers are instructed to only check this aspect. Some see this as a positive development, whereas others see it as only an excuse to publish more papers in order to make more profit, possibly at the expense of scientific rigour. A further consequence is that researchers are now flooded with requests to peer-review, leading to a crisis of the entire peer-review system.

Another development that can perhaps be questioned is the emergence of a score of OA journals tied to existing traditional journals with a strong reputation. A manuscript may be rejected by the traditional journal but instead referred to one of its associated more specialised OA journals, with the offer to transfer reviews and a strong chance of publication – given that an editor has already deemed the manuscript worthy based on the reviews. The tempting offer is often accepted, and the journal receives the publication fees. One problematic aspect of this procedure is that the 'trademarks' of highly rated journals such as *Nature* can be used to give sometimes misplaced credibility to lesser journals in the same publishing group (in this case, Springer Nature). It is not uncommon to see claims in social media hallowing a paper 'published in *Nature*' which is actually published in a highly specialised and very low impact journal but sharing part of the web link with *Nature*.

8.1.3.4 Submitting the Paper

Once the journal has been chosen, taking the above considerations into account, it is time to submit. This is done online at the journal's homepage, using a system which is typically shared among journals in a publishing group such as Springer Nature, Elsevier or Wiley, and involves not only uploading the text, figures and tables (formatted according to the journal's instructions), but also a cover letter to the editors and much other information. This can involve, for instance, short statements about how the results constitute progress in the field, about how the data will be made accessible, about ethical considerations and conflicts of interest or a description of how each author contributed to the manuscript. Sometimes authors are asked to suggest suitable editors, peer reviewers and so on. If not well prepared in advance, the entire procedure can easily take days to complete.

8.1.3.5 Editors and Editorial Boards

The editors of scientific journals are not paid administrative staff but scientists who do this for free in order to assist the peer review system and because it is a career merit to be selected for editorship at a respected journal. The predatory and grey zone OA journals exploit this aspect of the publication process, inviting scientists to their editorial boards to boost their own credibility.

For the peer review process to work as intended, editors must be scientists with some experience, because they make three important decisions: first, whether the paper should be sent for peer review at all or rejected outright; second, the selection of peer reviewers; and third, the final decision whether to accept the paper for publication or not. There is typically an editor-in-chief who first distributes the papers to an appropriate member of the editorial board (one with some expertise in the subject matter), who handles the paper from thereon.

A fraction of submitted papers are always rejected outright. This could be because they are not a good fit for the scope of the journal (wrong research programme in the sense of Section 5.4.1) or because it is obvious already to the editor that the manuscript is not of sufficiently good quality or is not novel and interesting enough for the standards of the journal. Highly rated journals such as *Nature* or *Science* in fact reject most papers already at this stage.

8.1.3.6 Peer Review

Papers that have passed the first hurdle are sent out for peer review, often to at least two reviewers. Given that many in academia are stressed for time and the flood of peer-review requests that they receive, it can take a while for the editor to find willing reviewers. The reviewers are then given a couple of weeks to log into the publisher's online system to get access to the paper, read it (ideally very thoroughly) and then log back into the system to provide a recommendation regarding whether the paper should be accepted, revised or rejected, along with a set of comments that can be both general and very detailed.

8.1.3.7 Editorial Decisions

Following peer review, it is up to the editor handling the paper to make a decision (or suggest a decision to the editor-in-chief) based on the recommendations and the editor's own reading of the manuscript. Possible decisions include: acceptance without revision, major revision, minor revision, rejection without prejudice or outright rejection.

Acceptance without revision is very unusual but happens occasionally. A *major revision* could include a thorough rewrite and/or additional analyses of the data, and for this reason, the editor will state in their decision letter that eventual publication is contingent on the outcome of the revision and not in any way guaranteed. A *minor revision* just entails correcting some small flaws and is often seen as a provisional acceptance. *Rejection without prejudice* (or 'reject and resubmit') is similar to a major revision, in that it is possible to submit the manuscript again, but it constitutes an even clearer signal from the editor that the paper may in the end not be accepted. It could, for instance, be that a new statistical analysis of the data is called for by the editor, and acceptance is dependent on the outcome of this analysis. If the manuscript is resubmitted, it will be handled as a new submission in the online system, rather than as a revised version of the original. In this situation, the authors will have to make a decision on whether it is better to instead try for another journal. Finally, an outright *rejection* means that a new version will not be considered. The authors will have to look elsewhere or decide to give up on publication if the manuscript cannot be improved, perhaps by adding more data or framing the results in a way that is more interesting and/or convincing.

8.1.3.8 Revisions

Revising the manuscript entails carefully assessing each comment from the reviewers and the editor, making changes and writing a response letter describing in some detail how each comment has been handled. It could be that there are unclear portions of the manuscript that need to be clarified, and in that case, changes are made to the text and the authors write a comment to that effect. If outright changes or additional analyses of data are suggested, the authors either make the requested changes/analyses or argue in the response letter why changes should, in fact, not be done or why the analyses are not necessary. For successful publication, it is usually a good strategy to do everything requested, but it happens that from the authors' better knowledge of the circumstances of the study and its logic, it is clear that the changes would be improper or not improve the manuscript, or that the suggested analyses would not be appropriate.

Once done with the revision, the new version is uploaded in the online system together with the response letter, and the authors wait for a new decision from the editor, often after a new round of peer review that may or may not involve the original reviewers. The hope is that the final decision will be one of accepting the manuscript for publication.

8.1.3.9 Publication

Once the publication is accepted, there are often some technical issues to deal with still, such as ensuring that figures and tables follow the standards of the journal or formatting the reference list exactly to the journal format. In journals that are not pure open access but have the opportunity to make individual articles OA ('hybrid' journals), a decision has to be made whether to pay the fee to do this. Once everything is in order, the publication will often appear on the journal homepage within days or weeks. In pure OA journals, this is the date of publication, but in traditional journals, it can be an 'online early' version, with the actual date of publication being sometimes months later, corresponding to its inclusion in a printed issue of the journal. This can lead to some confusion regarding the correct year of publication when the manuscript is cited.

8.1.3.10 Corrections/Retractions

As we have pointed out throughout this book, one of the strengths of science is that it explicitly recognised that results are fallible and theory-dependent. Because of this, it is sometimes necessary to issue a correction when a flaw has been found after publication. If the flaw is great enough that the conclusions can be questioned, the publication can even be retracted. This can happen when critical assumptions behind an experimental set-up have been found to be in error, or if scientific misconduct such as the manipulation of data has been found. Retractions can take place either at the request of the authors or following an editorial decision by the journal.

8.1.4 Finding Publications

Scientific publications can often be found via a simple web search on the topic of interest, with a link to online publications or to a PDF version. However, for a more systematic investigation of the state of knowledge on the topic, you need a database. Currently, the most readily accessible database is Google Scholar, where you can type keywords or author names in the search field and see what comes up. The lists of publications can be sorted by relevance or by date. Be aware, however, that this is not a curated database, meaning that it includes publications that are not peer-reviewed, such as pre-prints and theses. Another option is to ask an AI tool for a list of publications on a topic, but at least at the time of writing this list must subsequently be thoroughly checked for accuracy.

Some readers of this book might not get further than this, but students and staff at universities and other research institutes typically have full access (via subscriptions) to curated commercial scientific databases with more structured search options, such as *Web of Science* (run by the analytics company Clarivate) or *Scopus* (run by Elsevier). The Search page in such databases gives many options. In most cases, you would initially be looking for documents on a certain topic, and if so, you just enter words describing the topic in the appropriate field. If you have other information about a certain document that you are looking for, such as an author's name or the year of publication, you can add these and perform the search.

The next page gives the search results in the form of a list of articles fitting the search criteria. These can be sorted in different ways. You might want to sort the list with the newest or most highly cited articles first, or perhaps with the oldest first if you want to unravel the history of the research subject from the beginning.

Once you have found an article that you want to start with, you can click on the title to get to the page for that particular publication. You can read the Abstract and most often also access the full text. Old or specialised publications may not always be available in full, but they can sometimes still be found as a PDF via a web search on the title.

After having assimilated the publication using the tips in Section 8.1.2, there are several ways to explore the topic further. First, the reference list may show earlier articles of interest on the topic. Second, unless the publication is very new, it should have been cited by later papers, and they are an important source for more up-to-date information on the topic as well as to see how the publication has been judged by peers and how well its results have stood up to later investigations.

8.2 Citations and Impact Factors: Bibliometrics

Bibliometrics is a term for using statistical or mathematical methods to analyse books, articles and other publications. In Academia, it typically refers to various methods for counting citations of papers and how these counts are used to judge the career achievements of scientists and the impact of scientific journals and institutions. Counting citations may seem like a strange practice, but it makes some sense if it is recognised that having a publication cited is proof that it has not only been read by other scientists but also made some sort of impact on the 'consensus of rational opinion'. A citation of a scientific work generally means that another scientist has used it to back up their statements; in other words, it has been seen as part of the background knowledge on the subject. The citation could also refer to, for instance, a new method, which is also a valuable contribution to the research programme in question.

The use of bibliometrics is often criticised, rightly so if it is used too simplistically when, for instance, judging scientists for recruiting or funding. Indeed, there are several broad initiatives with the aim to improve merit judgement and make it more qualitative and dependent on thorough peer review rather than tabulation of numbers. One prominent example is the Coalition for Advancing Research Assessment (CoARA), signed by many universities and other organisations across the world. However, it is hard to deny that such metrics can carry valuable information, for instance, to determine whether a publication on a topic that you are interested in seems like a trustworthy source of information. As such, they are highly relevant for the web of trust in science.

8.2.1 Author Citation Metrics

One step in assessing a publication might be to look up the citation metrics of the authors. Although there are many caveats and complications, a well-cited researcher should, on average, be worthier of trust than one who does not seem to be recognised much by peers. There are many ways to count citations from an author, including the total number of citations, the number of publications with at least 10 citations and so on. Many of these metrics can be found in databases such as *Web of Science*, for instance, by doing a search for the publications of an author and clicking 'Citation Report', or by using *Google Scholar* to find an author profile listing the metrics.

8.2.1.1 The H-Index

The total number of citations can be misleading when judging the impact of a scientist, since they can be a co-author on a single paper with a high number of citations (perhaps a paper describing a popular method), with the remaining publications being virtually ignored. For this reason, the popular *H-index* (or Hirsch index, from the scientist who suggested it) was developed. It is intended to capture both scientific productivity and citation impact; a scientist having an H-index of X has published at least X number of articles, each having received at least X citations. This means that the career stages of scientists have to be taken into account when comparing their impact. Everybody starts with an H-index of 1 once they have been first cited, but a scientist with an H-index of 20 (at least 20 publications cited at least 20 times) after a 20-year career has, in this particular sense, performed better than one with an H-index of 5 at the same stage.

Note that a single highly cited paper does not much affect the H-index: having 1 paper with 1000 citations and 50 with 1 citation each means an H-index of only 1. In order to increase this index to 10, it is not enough that 1 of those 50 publications get 10 citations, you would need 10 papers in total with at least 10 citations each. This characteristic of the H-index means that it is pretty robust to, for instance, the impact of self-citations. This hypothetical scientist could easily increase the index to 2 by citing 1 of their own 50 low-cited papers, but it gets progressively more difficult: even if they should cite 20 of the papers in their next publication (which the reviewers would likely react negatively to), the index would still be only 2. Nevertheless, the index is often reported with self-citations excluded, since they can have an effect, particularly at early career stages when the indices are at low values.

Importantly, the H-index for a given author will differ depending on the database used to calculate it. The H-index in Web of Science, for instance, only counts citations from and to publications in journals (or book chapters, in some editions of Web of Science) indexed in this particular database, and is thus more conservative than the index given in Google Scholar. The latter uses an algorithm to find citations online, and this could include not only citations from publications in more specialised or otherwise obscure journals but also from PhD theses and so on. The result is a more generous score of the H-index.

8.2.1.2 First and Last Authors

If there are many authors, the process of looking up their citation metrics can take some time, and in such cases, it can be worth looking up at least the first and last author. Traditions vary between scientific fields, but often the first author is the one who did much of the actual work and wrote at least the first draft of the paper. This could be a relatively junior scientist, such as a PhD student or a postdoc, which should be taken into account when judging, for example, the H-index. The last author is often a senior scientist, the group leader. This is typically also the person who designed and won the funding for the project. It is to be expected, then, that the last author has more citations to their work than the first author. However, in some fields, the project leader is the first author at least if they actually wrote most of the paper, which means first authors should always be checked out as well. A high H-index is no guarantee that the science is good and vice versa, but it can be one of several indicators worth looking into.

8.2.2 Journal Citation Metrics

Citations are also counted for journals, following a similar rationale as for author metrics. A journal with many highly cited publications can be said to have more scientific impact than one whose publications are mostly ignored.

8.2.2.1 Impact Factors

The most commonly used journal metric is the *impact factor* (IF), which is simply the average number of citations for the last two years' publications in the journal. This metric can be found in another Clarivate product: the *Journal Citation Reports* database, which can be accessed from the Web of Science portal. As for the Web of Science H-index, only citations from journals indexed in the database contribute to the impact factor.

There is a hierarchy of sorts among scientific journals, from the few with very high impact factors at the top to the many with low impact factors. Not surprisingly, this hierarchy to a large degree follows the gradient from high-impact journals with broad scientific audiences (*Nature, Science*) to journals with very specialised audiences (see Section 8.1.3.2). Within a research field there is a similar gradient, from, for example, broad journals on ecology such as *Ecology Letters* (IF in 2023 7.6) or *Ecology* (IF 4.4) over the more specialised *Behavioral Ecology* (IF 2.5) or *Aquatic Ecology* (IF 1.7) to even more narrow-topic journals such as *Arid Ecosystems* (IF 0.6) or *Applied Ecology and Environmental Research* (IF 0.8).

There is much more to the impact factor hierarchy than this gradient, however. For instance, journals publishing reviews tend to have high impact factors because they are often cited in the Introduction of papers as part of the section establishing the background knowledge in the subject, whereas journals mostly publishing local authors and mostly of interest to a local audience (e.g. *Russian Journal of Ecology*; IF 1.0) have lower impact factors. There is also a 'brand name' component, where journals such as *Nature Ecology and Evolution* (IF 14.1) score higher than journals aiming for a similarly broad audience but that lack the association with *Nature*, such as *Ecology and Evolution* (IF 2.3). The latter phenomenon is also part of a self-fulfilling and somewhat problematic aspect of impact factors, where authors choose to publish their best (and presumably most citation-worthy) work in high-impact journals, because such publications are seen as a stronger scientific merit.

As for author metrics, the impact factor of the journal is an indicator of whether a published scientific result is trustworthy and should be checked out, but it is certainly not a perfect one. For one thing, it is a journal metric and only an average, so the publication in question may not have been cited very much even if the journal has a high impact factor. Databases now include attempts to judge the number of citations of a given publication in relation to the average expected in the journal or scientific field. For another, there are large differences among scientific fields in the speed of citation, affecting the impact factor if it is calculated only over publications in the last two years. Journals in rapidly moving fields where mostly recent findings are cited for this reason tend to have higher impact factors than those in fields where decades-old publications are still very much cited. This means that impact factors are best compared within a given field.

8.2.2.2 Alternative Journal Metrics

Several alternative ways of judging the quality of journals and other scientific publishers have been proposed, but here we will only mention the 'Norwegian list' (or more correctly the 'Norwegian register for scientific journals, series and publishers'), which is of interest because it is much more inclusive than the *Journal Citation Reports* database. Not only journals can be included, but also 'Series' and 'Publishers' (that publish scientific books). In some fields, especially outside of the natural sciences, it is common to publish results in a book – often in an edited series of books on a given topic – rather than in a journal article. Rather than attempting to give each source of publications an exact metric like the impact factor, the Norwegian system is simply an index of sources that are deemed to be scientific, at two levels. Level 1 is all those qualifying as scientific channels, whereas only the 'leading' channels in a given scientific field qualify for Level 2.

In order for a publication channel to qualify for Level 1, it is, for instance, necessary to have editors and a system for peer review. Authorship also has to be national or international, not from a single institution. Level 2 is restricted to the most respected channels, together publishing no more than 20% of the world's publications in a given field. Inclusion at this higher level is determined by how the channels are viewed by and used by scientists in the field, rather than by any particular metric. When evaluating scientists, a higher point can be given to publications in these leading channels.

8.2.3 University Rankings and Research Evaluations

Finally, it should be mentioned that (like it or not) citation metrics are also an important component in some of the systems for judging the quality of research and for ranking universities and other research institutions. In many countries, bibliometrics are used at the national level in quality assurance systems and for prioritising among research fields and research institutions when it comes to governmental grants.

Well-known systems for university rankings include the Chinese 'Shanghai ranking' (or more correctly the Academic Ranking of World Universities). Among the criteria used for this ranking, besides, for example, Nobel Prizes and similar awards are the number of highly cited researchers, the number of papers published in *Nature* and *Science* and the number of published papers indexed in major citation indices. Citations are also one of the components used to assess research impact in the World University Ranking performed by Times Higher Education. Measures of research impact are then combined with measures of teaching excellence and impact towards sustainability goals to produce the final ranking of universities. Needless to say, systems for evaluating research or ranking universities are far from perfect, and the outcome consequently varies between different systems. Much criticism has consequently been levelled against such rankings. To the extent that the rankings are used by governmental or other agencies in funding decisions, they cannot, however, be entirely ignored.

9
Science in Practice: Data

What is a rational opinion?

9.1 Scientific Data in the Light of Philosophy

At this stage of our explorations of the philosophy of science, it might be useful to take a closer look at the use of data and statistics in science. Science is invariably based on some sort of data collection and further treatment of the data gathered. We would argue that the choice of which type of data is considered relevant and which type of statistics used to treat the data (or indeed if statistics is used at all) is highly reflective of which philosophy the scientist adheres to, although this may well not be a conscious choice by the scientist. For example, controlled experiments are of central importance in some research traditions but not others, and the statistics used will be very different for a scientist following the Falsificationist or Bayesian traditions. In the words of Lakatos, such decisions are examples of the tacit methodological choices that make up the positive heuristics of a research programme. In the words of Ziman, it is how a scientist expresses what is currently accepted as a rational opinion in the particular field of the scientist, in the hope of affecting the scientific consensus among their peers.

9.1.1 Data from Pure Observations

Even 'pure', unstructured observations can play an important role in science. In the early days of an emerging branch of science (the *pre-science* phase, in Kuhn's terminology), such observations will be comparatively more important. At this stage, there may be few established theories upon which to formulate hypotheses, meaning that a more empiricist tradition may prevail. The science of palaeontology was, for instance, initially founded on a plethora of discoveries that were not explicitly sought for. Palaeontologists today primarily focus on exploring geological sites known to have a high likelihood of containing well-preserved fossils from a time period of interest to the scientist in the light of existing knowledge, but still never know precisely what they will find. The discovery of the 'Flores Man' *Homo florensis* in 2003 is a prime example of a highly unexpected find, at a site already known to have high archaeological and paleontological potential, the Liang Bua cave.

An ornithologist may happen to observe a species of bird far away from its known area of distribution, resulting in a published note in an ornithology journal. This contributes to the knowledge regarding this species and, at some point in the future, could perhaps add a data point to a larger study of dispersal patterns or navigation in birds. An astronomer might observe a new bright light in the sky, discovering what is later determined to be a comet or asteroid that has not been observed previously. Similar more or less random observations can occur in all sciences, adding to knowledge. Importantly, as we have seen earlier in the book, at least some training in the scientific field is typically necessary to recognise the nature and importance of such observations, demonstrating that they are not infallible and not theory-free. In the case of the Flores Man, it was even initially controversial if this was indeed a new species or a pygmy or malformed *Homo sapiens*, and although this notion has by now been plausibly rejected, the precise relationship with other species of *Homo* is still not entirely clear at the time of writing.

Accidental observations can occasionally lead to major scientific advancements. Alexander Fleming famously discovered penicillin in 1928 without searching for an antibiotic. He was instead studying different growth cultures for a bacterium but recognised the importance of the observation that a fungus contamination in one of the cultures killed off the bacteria. Fleming published a report on the finding already in 1929 [28], but it took several years before there was any appreciation for Fleming's notion that penicillin could be of value for treating infections, and mass production did not happen until 1945. Other researchers first had to find ways to isolate the compound and produce quantities large enough to test it on animals, and Fleming's later Nobel Prize was shared with two of these researchers. This well illustrates the social aspect of science – the importance of presenting results to peers in a way that makes an impact on the current consensus and the need for a community of scientists working towards similar goals.

9.1.2 Data from Structured Observations

Observational studies that follow a well-thought-out structure are of great importance in most fields of science, in a tradition going back even to the early empiricists. In Chapter 2, we saw how Francis Bacon tabulated observations on the phenomenon of heat in a highly structured manner in order to come up with a general theory of heat. However, most modern scientific approaches to structured observations rather attempt to formulate hypotheses based on existing theory and test them by emulating experiments as much as possible in situations when actual experiments are not possible. The terminology varies between authors, but such structured observations include *natural experiments*, *quasi-experiments* and *pseudo-experiments*. Such approaches are used in an attempt to infer how a 'treatment' causes an effect, even though – in contrast to true experiments – it is not possible to randomise samples over the treatments and to control for all variation besides the treatments.

In medicine, for instance, it may be possible to do an experiment studying the effects of brain damage on cognition in rats (disregarding for the moment the ethics of such research) by causing brain-damage in a random half of the rats and testing them all in a maze. In order to do a similar study on humans, you would instead compare the cognitive skills of healthy humans with a sample of humans with naturally occurring brain damage [29]. This is as good as can be done, but it must be recognised that the two samples might well differ in other ways because individuals are not randomly assigned to the two groups. Perhaps some categories of humans are more prone to get brain damage in the first place than others, and this propensity correlates with cognitive skills. Such *confounding variables* can be controlled for statistically, but only to the extent that they are known and already thought to be of importance based on previous theory. The possibility always exists that there are important variables not controlled for, meaning that inferences of causality are tentative at best when true experiments are not possible.

Research on the positive and negative effects of different diets, physical activity, etc. on human health provides a similar example, where most of the studies are in the form of following a large group of people over long time periods (often referred to as cohort studies). The persons accepting to take part in the study complete a standardised questionnaire detailing their characteristics and habits at the onset of the study and perhaps at intervals during the study. Such datasets often result in several publications, focusing on different aspects of the diet or activity, investigating how they correlate with health outcomes such as diagnoses and mortality from heart disease or cancer. Attempts are made to control for a range of confounding variables such as smoking, income and educational level. Typically, the researchers are testing a particular hypothesis, but it is not uncommon that a more empiricist approach is employed, looking for any interesting patterns in the data that can later be tested in a more targeted way. This is similar to the novel approaches to 'Big data' or 'data-driven science' mentioned in Chapter 2, where huge datasets are analysed using a range of methods (including machine learning and AI). Such studies also vary regarding the degree of structure in the data analysed and the philosophical approach, with at least some researchers explicitly striving to go into the analysis without formulating hypotheses so as not to bias the findings. However, what type of data is included in the datasets to begin with will always, to some extent, be shaped by existing theory.

'Natural experiments' of various kinds are commonly used in scientific fields such as evolutionary biology, ecology, geology or astronomy, sampling data from existing variation in a structured manner. In evolutionary biology, comparative phylogenetic studies are an important part of the methods toolbox. Here, an effort is made to see if the characteristics of organisms inferred to have a similar history of evolutionary selection pressures (for instance, inhabiting similar habitats) are similar even when they are not closely related, suggesting that natural selection partly shaped these characteristics as adaptations to their environments. To detect if the characteristics have indeed changed over the course of evolution in response to the environment, these species will be compared to related species inhabiting different environments, thus emulating the 'treatments' in a real experiment. Striking demonstrations of this principle are provided by cases of convergent evolution, such as the similar shapes of sharks and dolphins, but today sophisticated methods exist to detect much more subtle patterns of evolutionary change while statistically controlling for the similarities caused by characteristics that were instead inherited from shared ancestors.

In ecology one might compare the communities of species inhabiting different types of environments by sampling from a number of coniferous forests in different localities and comparing them with samples from a number of deciduous forests. Ideally one would, to some extent, control for other factors that might affect the species composition by adding a pseudo-experimental structure: comparing several pairs of samples, each taken from a coniferous forest and a nearby deciduous forest. Similar strategies can be used in geology to find consistent characteristics of earth formations with a similar geological history or in astronomy regarding classes of celestial objects, such as types of stars or exoplanets.

9.1.3 Data from Experiments

In a simplistic sense of trial-and-error, experimenting has been a part of human culture ever since early humans tried out the best way to shape tools or catch prey, and some modern human activities, such as cooking or car repair involve experimenting of this kind, sometimes even highly structured ('quasi-science' in the terminology of Chapter 7). We also know that the ancient Greeks performed more scientific experiments, in the sense that they were aiming to find out the nature of the world and not just make a life for themselves. Empedocles is usually credited with the first known experiment of this kind (around 430 BC), demonstrating that air is a substance by showing that it can displace water. During mediaeval times, experimenting flourished in the Arab world, directed by several prominent scientists such as Avicenna (medicine) and Ibn al-Haytham (optics). Following enlightenment and the dawn of science in a more modern sense, researchers such as Galileo Galilei and Isaac Newton designed groundbreaking experiments testing theories in physics. Even an empiricist such as Francis Bacon performed many experiments as part of his information-gathering, for instance, heating different metals to see if they expanded. However, this was not done to test hypotheses but in order to gather good observations upon which to form a theory by induction (Chapter 2).

In science today, well-designed experiments testing explicit hypotheses are typically highlighted as the gold standard of scientific research. This is because in an ideal experimental situation – where you hold everything constant except for one factor that you change – it is possible to infer that it was this manipulated factor that caused an observed effect, and not something else. This is a scientific ideal handed down primarily from physics and chemistry, where it is possible to approach such a controlled situation to a much greater extent than in biology or sciences dealing with humans (more on this in Section 9.2). Even in physics and chemistry, however, experimenting does not always produce clear-cut results.

For one thing, it has been repeatedly demonstrated that it is hard to eliminate all subjectivity when recording and interpreting experimental results; it is easy to see what you want to see, that is, results consistent with the hypothesis that you favoured. For this reason, some sort of blinding should be used when possible, perhaps by analysing results without knowing which data is from which treatment or by letting a third-party record and analyse the results. Second, effects have to be recorded and measured in some way, and such measurements are never exact; there is always some error variation to be dealt with. Third, it is not an easy task to design an experiment to conclusively test a hypothesis, in such a way that a given experimental outcome undeniably supports or falsifies the hypothesis. This is because of the theory-dependence of the experimental set-up as well as of the outcome and analysis (see Section 5.4.2 on the 'New experimentalism' philosophy of science). Fourth, precisely because the experiment is a carefully controlled situation, it is not always clear how relevant the experimental results are in the natural world outside of the laboratory. Steps can be taken to vary more than one factor in the experiment in order to see how they interact, but this is still far from the complexity of the real world.

9.1.4 Models in Science

At this point, a mention should be made of the central importance of *models* in science. A model is a description and tentative explanation of a phenomenon in the universe, used by scientists to organise observations and thoughts and to communicate these ideas to other scientists. Models can be verbal, visual or mathematical – in other words, conceptual or quantitative – depending on the phenomenon and the degree of knowledge amassed on the topic. They can and should change when science advances, and alternative models can also be explored and tested. In this, models are closely related to scientific theories and to hypotheses derived from these theories. Models allow newly collected data to be placed in such theoretical contexts. Philosophers of science with a focus on science in practice have shown how the activity of modelling itself, not just the resulting models, helps scientific reasoning, theory construction, experimental design and conceptual innovation [30].

Widely accepted, particularly fundamental models can form part of a paradigm or the hard core of a research programme (Chapter 5). The paradigmatic idea that matter is ultimately made up of small parts that we can call atoms can, for instance, be seen as a model of one of the most fundamental aspects of the universe, going back to the ancient Greeks (see also Chapter 2 on logical positivism and antirealism). Visual models of this phenomenon might depict atoms as dots on a whiteboard or balls that can be put together to form molecules. With increasing knowledge these simple models were replaced with models depicting atoms as having a positively charged nucleus surrounded by a cloud of negatively charged electrons orbiting the nucleus, and then with the model proposed by the Danish physicist Niels Bohr in 1913. Bohr wanted to understand why atoms only emit light of specific frequencies and incorporated quantum theory in an attempt to propose that electrons can only exist in one of a set of allowed orbits, with light being emitted when an electron transfers to an orbit closer to the nucleus. Hypotheses derived from this theory were tested experimentally already in 1914, with results supporting the theory. This model goes a long way to explain the properties of elements and why atoms of certain elements have a propensity to bind with atoms of other specific elements to form molecules and others not. Modern models of atoms are highly sophisticated and quantum-mechanical, attempting to explain the properties and behaviour of matter and light by incorporating subatomic esoteric particles such as quarks, gluons and the Higgs particle. The latter was predicted by the so-called Standard Model of particle physics already in 1964, but it was not until 2012 that a particle with the expected properties could be convincingly demonstrated to exist.

Notably, even though we now have a deeper understanding of what an atom actually is, the simpler models are still useful in many contexts, such as standard chemistry. This illustrates a general principle: the most complex model is not always what we need. Following the principle of parsimony (Chapter 2), models should not be made more complicated by adding assumptions until this is needed, as it was for the model of the atom in order to explain quantum phenomena but not before. Simple models also often have the advantage of being more generally applicable, if only to some extent, whereas a more complex model can give a more precise description and explanation of a specific case. The matter of how complex models should be has for this reason been much debated by philosophers and scientists, particularly in ecology [31], and we side with those who advocate the pluralist approach suggested by Richard Levins [32] – the choice of model complexity should depend on whether the aim is that it should apply to many systems or whether a realistic and precise output is needed.

A related concept is the extensive use of *model organisms* in biology, where a few particularly well-studied species (such as the fruit fly or the rat) serve as models for how also other species function, including humans. This is done because it is not possible to study the genetics, developmental biology and physiology of every species in similar detail. Model organisms are also used for experiments that would not be feasible or ethical on humans. The practice is based on the paradigm of common descent – many functions are inherited across speciation events and thus conserved over evolutionary time. The expectations from model organisms will not always hold, because the species have in fact diverged since they shared a common ancestor, so to minimise this problem a model is used that is as closely related as possible to the organism currently under study.

9.2 Dealing with Variation

Why is statistics used at all in science? This is done to deal with variation among samples or experimental results in order to establish how confident we can be in the results in the face of this variation. In physics and chemistry, statistics primarily deal with error variation – variation in the precise measurements in an experiment. This is especially crucial in fields such as analytical chemistry, dealing with developing methods to determine which elements and chemicals are present in a sample and in which amounts. Statistical methods are also used to study the behaviour of aggregates of atoms and molecules, going back at least to Daniel Bernoulli and his kinetic theory of gases in 1738, and in *chemometrics*, the data-driven analysis of complex chemical systems. Research on the living world, such as biology, medicine, psychology, and other subjects dealing with humans, is confronted by *biological variation*, most often necessitating the use of statistics to analyse the results from observations or experiments.

9.2.1 Variation and Statistics

Say, for example, that we try to establish that males and females of a given species differ in size. Not all individuals of the same sex and species have the same size, and we must confront this variation to see if we can come to the conclusion that the sexes differ in size, while taking this variation into account. What we then do is to take a sample of several individuals of each sex and use a statistical test such as the *t*-test to infer whether the sexes have different size distributions. If the distributions of sizes in the total population were actually equal for the sexes, there would be a low probability that you would find very different sizes in a big enough sample of individuals. By the same logic, if you do find a difference in such a sample, the null hypothesis of equal size can reasonably be rejected. The result that the sexes (probably) differ is said to be statistically significant (Figure 9.1).

We will return to the concept of null hypotheses in the next section, but let us first deal with the fundamental aspect of biological variation. Variation among individuals in all kinds of traits, both within and between the sexes, is found in all species of living organisms. Individual variation can have a genetic basis, meaning that variation can be found also among groups of related individuals, with similar genotypes and thus similar inherited traits within groups but more different DNA sequences between groups. Variation can also be directly influenced by the individual's environment, in particular during growth and development, thus causing variation among groups of individuals raised in different environments. This can be extended to variation among populations within the same species, inhabiting different geographical sites (differing in their environments) and having differences in genotypes both due to random genetic processes and due to natural selection adapting the populations to the local environments. We can stop there, hopefully having illustrated how the presence of biological variation means that some sort of statistical treatment of scientific data is almost always necessary in biology and how conclusions can always be only probabilistic since we can never sample the whole species. Moreover, the existence of such variation means that experiments can never be exactly repeated in biology, medicine, psychology or social sciences (because there will always be at least some biological differences between the experiments). Although one can strive for as much repeatability as possible, this is likely part of the reason for what has been called the *replication crisis*, the growing awareness that many classic experiments, especially in psychology and medicine have a low repeatability.

Similar considerations would to some extent apply to scientific fields dealing with variation of other kinds. In the geosciences there is variation among sites and among samples of whatever entity is studied. No piece of mineral has the exact same composition as another and no location is exactly the same as another. Statistics are for this reason heavily used in geology. Such *geostatistics* was originally used to characterise the geological properties of an area to, for instance, find areas suitable for mining or oil drilling, but there is today a wide use of geostatistics in all fields dealing with spatial variation. Likewise, in astronomy no planet, star or galaxy is exactly the same as another, and statistical methods have to be used to identify what is observed. All of these fields of research are similar in that the variation is due to the precise history of each of the entities studied, but biology and studies on humans stand out in that much of the variation can be ascribed to history in terms of biological relatedness, the degree of common ancestry. Much of physics and chemistry is quite different from all of these fields, in that with some exceptions (such as variation among isotopes) a reasonable assumption can be made that the entities studied are basically invariable: one atom of gold or one molecule of water is the same as another.

Fundamental differences between fields of research like these show the need for different methodology in different fields, including if and how statistics are used. This is one example of what Kuhn meant when outlining how a paradigm in a phase of normal science must develop methods within the paradigm that improve the fit between theory and actual results, or what Lakatos termed positive and negative heuristics in a research programme. Here, we do not have the intention to go deeply into statistical methods, but we wish to point out some issues where philosophical decisions must be taken, rather than simply trusting the output from a statistics programme. This applies also to how you interpret statistical results presented by others. Often not enough thought is given to the philosophical aspects of statistical treatments. Rather, scientists tend to follow what others do in their paradigm/ research programme, and scientists in other fields or non-scientists tend to trust the results presented – by necessity, since it is not easy to evaluate statistical tests on types of data outside of one's own expertise. However, we believe much can still be gained by a higher awareness of how philosophy enters into statistics.

9.2.2 Falsification and the Null Hypothesis

In Chapter 4, we introduced the Duhem–Quine thesis as a serious problem for falsificationism. In brief, since observations are theory-dependent, there is nothing that allows us to say that an apparent falsification should be blamed on the hypothesis. It could just as well be the observation itself, or the assumptions inherent in the test situation, that should be falsified. In Chapter 5, we showed how the larger theoretical frameworks that scientists are working within can provide guidelines for such methodological decisions so that scientists can proceed in a constructive way rather than losing themselves in this maze of assumptions. However, there is another way to deal with the Duhem–Quine thesis, based on the relationship between complementary hypotheses. When we wrote in Chapter 4 that if you have a very precise hypothesis, such as 'there should be an object in that exact place in space' and actually find an object there, it would feel like a confirmation of the hypothesis, even though one *could* also phrase it as if you have falsified the hypothesis 'there is no object in that place'. These two hypotheses are non-overlapping; they cannot both be true at the same time and they cannot both be false at the same time. It turns out that scientists have sneakily found a workaround to the Duhem–Quine thesis by taking advantage of the inherent asymmetry between such complementary hypotheses, and this is typically done using statistics.

The complementary hypothesis to your main hypothesis is often called the *null hypothesis*. The concept of the null hypothesis can have slightly different meanings, but the most common is the one advocated by the statistician Ronald Fisher: in short, it is a formulation of the expected outcome of an experiment if the main hypothesis is false. Again, the two hypotheses have to be formulated in such a way that both of them cannot be true at the same time.

If you are familiar with classical statistical hypothesis tests (often called Fisherian statistics, after the aforementioned Ronald Fisher), you know that they are actually designed to test the probability that the null hypothesis is false not the main hypothesis (for an example see Figure 9.1). Perhaps we have found a promising drug and want to test if it really has a desired effect against a specific illness. We can then set up a clinical study to test the prediction that patients treated with the drug should show a speedier recovery than those given a placebo. The null hypothesis – the base expectation if the main hypothesis is wrong – would be that there is no difference between the two groups. The

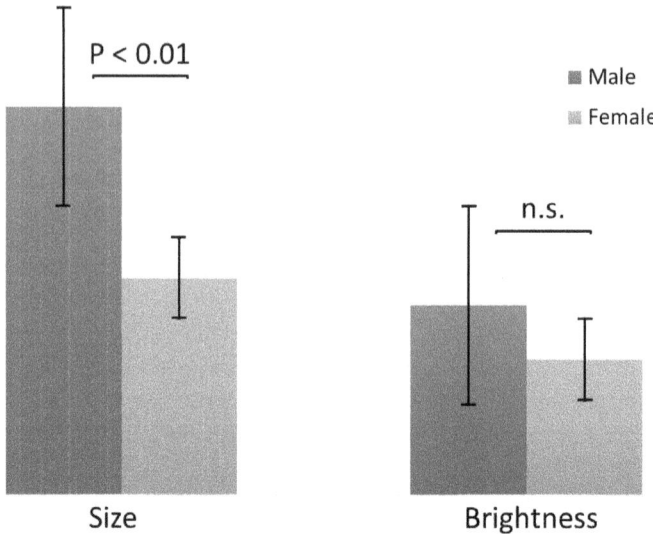

Figure 9.1 Results of Fisherian statistical tests (such as *t*-tests) of the null hypotheses that males and females are the same size and have equally bright colouration in an imaginary organism. Staples show averages, and vertical bars show standard errors (a measure of variation among the samples). The null hypothesis of equal measures is rejected for size (less than 1% probability) but not for brightness (n.s. = not significant, i.e. >0.05% probability). The overlapping standard errors in the latter case indicate that the sex difference found may have been due to chance when sampling from the population.

statistical test then tests how likely it would be to get the observed result if the null hypothesis is true and there is no effect of the drug on recovery. If this probability – the *p-value* – is low, we can say that either the null hypothesis is false, or something improbable has happened (we will return to the *p-value* in Section 9.2.3). If the null hypothesis can thus be reasonably rejected (falsified), scientists typically draw the conclusion that the main hypothesis is likely to be correct. How can this be justified? It would seem that what we learnt in Chapter 4 about the Duhem–Quine thesis should make us suspicious of such a conclusion. If we cannot falsify our main hypothesis, what makes us think that we can falsify the null hypothesis? And even if we could, what is there to say that our main hypothesis is responsible for that outcome, and not some other, yet unformulated, hypothesis?

This is where the trick comes in. The Duhem–Quine thesis doesn't really tell us that we cannot falsify, but that we cannot say what we have falsified. We saw in Chapter 4 that this would be a big problem if we were to try to falsify our main hypothesis, because we are told to be precise when formulating it. Ideally, it should be the only possible explanation of the predicted outcome given the experimental set-up. But that means that if we don't get the predicted outcome, we haven't learnt much, since that outcome could be caused by any number of things. On the other hand, this is exactly why by falsifying the null hypothesis, we have learnt something. If our hypothesis was indeed the only thing that could explain the predicted outcome, it doesn't matter so much that we logically cannot say exactly what caused the falsification of the null hypothesis, because it should be a highly unlikely outcome in the absence of our hypothesis. Hence, if we do get this outcome, the hypothesis is likely to be true, as well as all the underlying assumptions we used to make the prediction. Of course, it is virtually impossible to approach this level of precision in the real world, but you can try to get as close as you can. This is one reason why experimental design is so important and so difficult. The better we are at formulating precise hypotheses and designing our experiments to test predictions derived from them, the more confident we can be that the falsification of the null hypothesis really does imply that our main hypothesis is correct.

It is still important to remember that null-hypothesis testing can never give us a definitive answer. This is not only because of the inherent uncertainty inherent in the *p-value* (it is a *probability*, after all), but also because of the logic of the test situation. Researchers must always be aware that it is possible that falsification of the null hypothesis is caused by some alternative hypothesis that we have not thought of, or some error in the experimental set-up. The ideal situation just described – where rejection of the null hypothesis is highly likely to imply that the main hypothesis is true – is probably quite unusual. Typically, several complementary experiments are needed to be reasonably confident in a hypothesis, using different set-ups and different auxiliary assumptions. Still, the statistical approach of null-hypothesis testing has proven to be relatively robust and is one reason why falsificationism is still so prevalent in scientific practice, despite its shortcomings.

9.2.3 Significance Thresholds

Let us now return to the *p-value*, as it is not an entirely straightforward concept. The most commonly used threshold for significance is $p < 0.05$. In the simple example of comparing two categories, as in the example of males and females in Figure 5.1, this should be read as a probability of less than 5% to get the difference found between the sexes (or an even greater difference) when sampling individuals randomly from the two categories, if the sexes in fact do not differ. This threshold is, however, pretty arbitrary, it is a philosophical choice. You will get a significant difference just by chance one time out of 20 (equal to five in a hundred, or 5%) when sampling randomly even if there is no difference. The 5% cut-off is an attempt to balance the risk of such 'false positives' against the risk of concluding that there is no difference, even though there is. Sometimes a stricter threshold for significance is chosen, such as $p < 0.01$, when it is important to rule out more false positives. If you want to judge whether a new medical drug is effective, you may not want a 1 in 20 chance of administering an ineffective drug with potential side effects. Conversely, you could choose a less strict threshold, such as $p < 0.1$, if you are just exploring data in a search for potential differences that perhaps can be further explored in further studies. The number of statistical tests performed on a dataset is another issue. If 20 tests are done, you would expect that on average one of them becomes significant just by chance if the $p < 0.05$ threshold is used. There are ways to correct for this, lowering the threshold for significance in proportion to the number of tests. However, if you are in an exploratory phase you might, depending on your scientific philosophy, choose to consider all significant results as worthy of pursuing further.

A less strict threshold is also used in practice if you choose a 'one-tailed test' rather than a two-tailed test. A two-tailed test (which is typically used) looks for differences in any direction between the categories, but a one-tailed test considers differences only in the predicted direction. We might predict that males should be larger than females because males of the species compete violently for females and choose a one-tailed test. However, this practice is in most situations considered inappropriate, in particular if it is used just to more easily obtain a significant result. Furthermore, it might well be that females in fact turn out to be the larger sex, leading to a new theory for how size evolves in the species. Use of a one-tailed test would fail to pick up a significant difference in this unexpected direction and impede scientific progress.

There are different views of the significance thresholds in different research traditions. Scientists raised on falsificationism may argue that the chosen threshold must be strictly respected, so that it becomes logically possible to falsify the null hypothesis. With this view, a probability of 0.049 for the found difference means that the null hypothesis can be rejected and there is an actual difference, but a probability of 0.051 means that the null hypothesis survives falsification and still stands, so there is no difference. This view and similar fixation on significance thresholds in science may be part of the reason for the evident publication bias, where reported probabilities around 0.049 and just below are much more common in the literature than 0.051 and just above, and the same applies to lower thresholds such as $p < 0.01$ [33]. The extreme falsificationist stance regarding statistics seems to have become rarer over time, however, and it is now common to see a more nuanced discussion of statistical results, where a probability of, for instance, 0.055 would be presented as 'near-significant' and worthy of noting. This attitude may reflect a choice to acknowledge that probabilities are not set in stone and can differ even when an experiment or other type of sampling is repeated. One could argue that since the *p-value* is a probability, it already contains the necessary information for evaluating the likelihood of a hypothesis being true or false; a significance threshold is only needed if falsification is a goal of science.

There are in fact several reasons why a probability of less than 5% should not be taken as establishing that there is a 'true' difference. One is the variation present in nature, meaning that probabilities will vary with every attempt to sample from the categories being compared, sometimes even producing a false positive. Another is the theory-dependence of observations elaborated upon in earlier chapters, meaning that the sampled observations can be in some sense faulty. Moreover, statistical tests of this type make the critical assumption that the sampling is done in a random manner from the categories being compared. This assumption is easily violated, if there is a confounding factor present that biases the observations. When comparing males and females, as in Figure 5.1, for instance, what if only the largest males were sampled because the smaller ones are outcompeted and leave the area? For every comparison between categories, it is easy to imagine a whole list of potential confounding factors, and as Deborah Mayo pointed out, it is the job of the scientist to make every effort to exclude or control for these when designing experiments. This is even more crucial when a non-experimental method is used, such as a longitudinal study following a cohort of humans over the years. In such studies, statistical tests will attempt to control for variation in, for example, income or smoking habits, but the possibility always remains that yet another factor has not even been recorded and thus cannot be controlled for.

9.2.4 Predicted Results and Chance Findings

Another aspect where philosophy of science enters into statistics is the difference between predicted significant results and statistically significant results found more by chance when analysing data. An empiricist take could be to see all results as equally important and believable, and as noted in Chapter 2, 'data-driven science' is emerging as a new research tradition with this basic 'unbiased' approach. Others would argue that predicted results are, as a general rule, more trustworthy, because evidently there was a reason for the prediction. In the original empiricist tradition this is actually an integral part of the philosophy; predictions are made by deduction from already acknowledged knowledge. Also, when following hypothesis-testing philosophies such as falsificationism, predictions can be said to be grounded in the current background knowledge in the field of research (although a hypothesis should conflict at least somewhat with established knowledge in order to be novel and interesting to test). A result in accordance with predictions is thus more believable than an unexpected result and at the same time also gives more support to the parts of established knowledge that were used to make the prediction. The use of so-called posthoc tests to explore different aspects of the data after the main test has been performed is something of a middle ground, depending on whether the tests were planned ahead of the analysis (i.e. testing hypotheses regarding the details of factors contributing to the main result) or not. In the latter case, there is an increased risk that significant results are just spurious, created by some chance effect or confounding factor.

As we have seen, a statistically significant result predicted by a hypothesis (null hypothesis falsified) still does not prove that the hypothesis is correct, it only lends some more support to it. There could be other reasons for the result, such as a different theory that could also have predicted it or a failure to control for a confounding factor. Anecdotally, we have noticed that publications of questionable scientific quality tend to more often report probabilities much lower than the standard thresholds of $p < 0.05$, $p < 0.01$ and $p < 0.001$, such as $p < 0.000001$. This could be seen as a warning signal to look for a confounding factor or flaw in the sampling design that better explains the result. More generally, the mathematical reasons for significant results should always be scrutinised in order to judge the support for the presented hypothesis. With very large sample sizes it is easy to find highly significant results, but they may still explain very little of the variation.

So far, we have used statistical tests regarding differences between categories as our main example, but all of the considerations given in this chapter apply also to, for instance, tests of correlations between two variables, often in an even stronger sense. In a hypothesis-testing tradition we might have the hypothesis that the factor A is involved in causing variation in trait B, and test the prediction that if this is the case, variation in the two variables should be significantly correlated. If they are, this lends some support to the hypothesis of causality, but only to the extent that it was a reasonable hypothesis well-grounded in background knowledge in the first place. Conversely, looking for correlations between variables without prior predictions is associated with a very high risk of spurious significant results because of confounding factors. To take a simple example, variation over the year in almost any environmental variable could be significantly correlated with many other environmental variables, without a causal relationship. Plants may wilt in the summer heat, and at the same time people get tanned, but one does not cause the other; sunlight is the confounding factor.

Concerns with the temptation to look for statistically significant patterns wherever they are found, coming up with untested ad hoc explanations for them in scientific publications, have led to a movement demanding that research plans and statistical tests should be *preregistered*, that is, submitted to an open registry before the study takes place. When doing so, it is possible to distinguish between exploratory research (which can generate hypotheses) and hypothesis-testing research, but the same data cannot be used to both generate and test hypotheses.

9.2.5 Alternative Hypotheses and Bayesian Statistics

In the examples given so far in this chapter, there is a single hypothesis being tested, but a choice can also be made to test two or more alternative hypotheses against each other – to see which one gets the strongest support from statistical tests. This can be a reasonable alternative, especially if several potential explanations for some observed pattern have been put forward in the scientific literature. Note, however, that such a practice clashes with strict falsificationism, since one of the hypotheses will always turn out to be best supported, even if the actual support is in fact pretty poor. Thus, there is no strong attempt at falsifying any of the alternative hypotheses. A similar practice is involved in model selection, where different attempts at modelling a complex system are tested against each other to see which one can best predict patterns without complicating the model too much by adding factors that do not improve the fit of the model. Null modelling is a more stringent variant of this exercise: to test a hypothesis, a null model is created which leaves out the focal mechanism of the hypothesis, and the match with data is checked. If the match with the null model is poor compared to that with a model based on the hypothesis, this gives some support to the hypothesis. If it is good, this is evidence against the hypothesis, because the focal mechanism is evidently not necessary to produce the data observed.[1]

Finally, Bayesian statistics should be mentioned as an alternative to traditional statistics, grounded in a different philosophy. Traditional statistics is sometimes called the *frequentist* approach because it is based on repeated events such as sampling many males and females, counting up the frequency of males being larger than females (or vice versa) to see if this frequency deviates from the null hypothesis of equal size. Another simple example would be flipping a coin, with the expectation of equal numbers of heads and tails,

---

[1] For an interesting discussion of the appropriate null hypothesis and null model for testing alternative hypotheses regarding how mate choice may cause sexual dimorphism in colour and displays, see Kovaka [34].

given a large enough number of repetitions. In other words, a 50% frequency of heads has the highest probability, and the hypothesis that the frequency should approach 50% when the sample size increases is fixed and does not change depending on the outcome of coin flips. In Chapter 4, we encountered the idea from Bayesianism that probabilities should not be decided beforehand in this way but rather be continuously updated as new information comes in. If the frequency of heads actually approaches 90% as we go on flipping, surely, we should move to the new hypothesis that the coin is loaded to produce this result, rather than just falsifying the null hypothesis? In traditional hypothesis testing grounded in falsificationism this is not allowed, because it would entail using the same data to generate and test the new hypothesis.

Various forms of Bayesian statistics, including for the purpose of model selection, have become increasingly popular over recent years. We would argue that the frequentist and Bayesian approaches are complementary. If the purpose is to find the best description of a specific case, such as this particular coin, this particular geological area or the nutrient flow in this particular lake, Bayesian and similar approaches to statistics and model selection make good sense. However, when we search for more general patterns, such as what causes sex differences in the size of animals or how nutrients flow in lakes in general, the frequentist approach to hypothesis testing is still the most powerful.

9.3 Reviews and Meta-Analyses

Besides 'primary' publications (i.e. the ones containing novel results), another form of scientific publication is the *review*, where researchers report on the current knowledge on a particular topic. Ambitious reviews contain a synthesis that goes beyond just summarising the results. A further step towards providing something akin to new results is the *systematic review*, where the topic and type of studies to be included are strictly defined and results critically analysed, perhaps reaching conclusions not evident from the individual studies.

The *meta-analysis* is a form of systematic review synthesising quantitative data from several independent studies on a particular research question, using statistical methods to arrive at a single combined result. This is done by extracting effect sizes and measures of variance from each of the studies and computing a combined effect size which is then analysed statistically to see if it is significant. The combined result will be impacted more by studies with large sample sizes than by small studies, and more by studies with comparatively little variation among samples within categories. Sometimes only studies meeting some criteria of scientific standards are included. Meta-analyses are highly regarded in some fields, such as psychology, medicine and ecology, but they are not without challenges. These include the fact that studies vary in design and may not be simply added together, and a small study may even have been better designed than a large one. Publication bias is another problem; if non-significant results are left unreported a combined analysis may be misleading.

9.4 Combining Evidence: An Example

We will use the study of sexual size dimorphism in animals alluded to in this chapter (as a hypothetical) to illuminate how science in the real world most often proceeds not by spectacular discoveries but by evidence from different types of studies, accumulated over time.

In many animal taxa the female is the larger sex, probably because the number of offspring that can be produced by females is often more limited by size than is the case for males. There are some spectacular exceptions, notably in species where the males compete aggressively for access to females for mating, such as the walrus or gorilla. Such patterns, along with male weaponry and sexual differences in colourful displays, were noted by Darwin and led him to propose a theory of sexual selection to complement the theory of natural selection. Darwin wrote in his *Descent of Man* [35] that males are indeed larger than females in most mammal species. This statement was not really backed up by data, only by selected examples, but was accepted as fact for a long time. A number of studies of single species of mammals over the years found varying patterns, often with larger average sizes of males. To mention just one example, Deroucher and colleagues sampled all polar bears that they could find at an Arctic study site in an attempt to get a representative sample of the Barents Sea population [36]. They found that the sexes did not differ in size early in life, but the average size and mass of males was significantly larger already in year-old cubs.

More recent studies, however, have shown that in many or even most mammal species size dimorphism is not very pronounced, and larger females are a relatively common pattern as well [37, 38]. Still, when there *is* pronounced size dimorphism, males are typically larger, and comparative studies involving species with different mating systems lend support to Darwin's idea. In primates, for instance, sexual size dimorphism with larger males is most pronounced in species where one male can monopolise access to several females in a harem [39]. Some within-species studies have also found that sexual selection has not only shaped the patterns but is still occurring. In a study on fallow deer, larger males had higher mating success [40].

Genetic studies have been another important track. Darwin lacked a modern understanding of genetics, but Fisher outlined theoretically how divergent selection on the two sexes could accumulate genes with different effects in males and females, causing a character to become sexually dimorphic [41]. Russel Lande later developed detailed genetic models for how this process can occur, despite the strong genetic correlations expected between the sexes [42] (after all, most genes occur in both sexes), and Reeve and Fairbairn tested his models by using computer simulations and found that these genetic correlations seem to be less of a constraint on evolution than predicted by Lande [43].

Experiments to test such models are for practical reasons only possible in species with short generation times. Selection experiments have for this reason been performed mainly on fruit flies, with results that have by extension informed also theory on mammal genetics and evolution. In a rare mammal study, Nils Korkman in 1957 published the results of a selection experiment on mice, where he selected the largest males and smallest females in one strain and the opposite in another [44]. He saw a response to this selection after 10 generations and thus perhaps a genetic basis for sexual size dimorphism. However, males were still larger in both strains, and there was no replication of the strains, meaning that a spurious result cannot be ruled out.

A couple of recent meta-analyses have attempted to summarise where we now stand regarding Darwin's idea, using comparative studies from many animal taxa. This is not trivial, since the strength of sexual selection can only be assessed indirectly using proxies such as the mating system and variance in mating success. Janicke and Fromonteil searched the literature for studies estimating the opportunity for and/or strength of sexual selection in both males and females and extracted measures of sexual size dimorphism for these species from the literature [45]. They performed a meta-analysis of data from 59 species while controlling for the phylogenetic relatedness among the species. Results were mixed but did not show a strong link between sexual size dimorphism and estimates of sexual selection, possibly because of flaws in these estimates. In contrast, Winkler and colleagues did find a strong link in a quite different meta-analysis, this time not combining evidence from within-species studies but from 77 earlier comparative studies attempting to investigate correlations between sexual size dimorphism and estimates of sexual selection across species [46]. Most of the studies that were included concerned mammals or birds. Winkler and colleagues found that strong sexual selection from variance in mating success was typically correlated with male-biased size dimorphism, but with exceptions such as birds with aerial displays, where small males may be more agile. Furthermore, they noted that there may well be a bias in published comparative studies towards taxa showing clear male-biased size dimorphism.

In summary, the current consensus probably leans towards support of Darwin's original proposal, but there is still need for further research in order to understand the scope and limits of his theory. To what extent is size dimorphism instead explained by sexual selection on females? By other selection pressures altogether, differing between the sexes for other reasons? In this case falsifying the null hypothesis of equal size in wild populations is apparently not enough to prove the hypothesis that sexual selection on males is what causes males to be larger than females, because alternative explanations are possible. The simple logic of null-hypothesis testing (Section 9.2.2) does not really apply when we are trying to understand the outcome of a historical process by structured observations, in contrast to when it is possible to perform more strictly controlled experiments in physics, chemistry or even molecular biology. However, if the size dimorphism is predictable from theory and detailed knowledge of the mating system in the studied species, we can have some confidence in accepting the sexual selection hypothesis.

As this example should illustrate, single studies rarely deliver the truth by themselves, as they may contain flaws or only be able to address one aspect of the claim. Rather, it is the combined evidence from theory, observational studies and experiments that forms the current scientific consensus (more on this in Chapter 10). The consensus is also not static, it may well change with new evidence.

10
Science in Practice: Academia

What makes a competent researcher?

10.1 Academia and the Competent Researcher

In examining how the social definition of science can help us understand how science is done in practice, we have now reached its last component: the *competent researcher*. In Chapter 6, we already noted the importance of the academic system for maintaining the web of trust and scientific integrity but also noted its role in producing the scientists themselves. In this chapter, we take a closer look at this academic system and discuss how modern science is performed in practice in the world of Academia and elsewhere.

We do not wish to imply that the way science is currently performed is perfect. Most scientists would be able to point at features of the current system that are problematic, some of which will be noted in this chapter. Like any complex societal structure Academia also has a natural inertia and will often tend to lag behind rapid changes in society at large. However, for the most part it works reasonably well and at least provides a workable solution to the problems and needs we have outlined in this book.

10.1.1 The PhD Education

First, who is a 'competent researcher'? For science to work well as a public discussion of ideas in search of a consensus, not all opinions can be of equal weight. Science is not democratic in this sense; it is meritocratic. To be counted as a scientist worth listening to, you must know your subject: What is provisionally accepted as knowledge in the field? What are its main theoretical structures? What are its accepted methods? Only then can you hope to be taken seriously when proposing to change any aspect of the current consensus.

The simplest way of knowing who qualifies to take part in this discussion is by means of formal education in the subject, so you typically need a PhD or at least to be on your way to achieving this degree. To be accepted as a PhD student in the first place you need a diploma in the subject of your PhD or a closely related field, typically a master of science or equivalent. One of the reasons we have universities is to provide a form of education where scientists with PhDs take part in teaching already at the undergraduate level, thus increasing the chances that the contents of the classes reflect something of recent progress and ongoing controversies in the subject, not only long-established (or outdated) 'facts'. Ideally then, a student graduating from a university will be well prepared to take on a PhD education and start an academic career. In practice such opportunities are limited, not all students are well suited (or so inclined), and it is not easy to select the very best candidates. The best predictor may be not the grades from classes but rather the quality of the master's thesis (involving research with a degree of student independence) and recommendations from supervisors familiar with the process leading up to the thesis: how the student dealt with issues such as unforeseen complications, statistics and scientific writing.

The exact form of the PhD education varies among subjects and countries, but the similarities are probably stronger. There is some course work to further deepen the knowledge of the subject at large, but most importantly it involves a real scientific project. This project is led by a senior scientist acting as supervisor (sometimes referred to as advisor) but the PhD student is expected to contribute substantially and with increasing independence over time to the design of experiments/observations, analysis and writing, thus maturing to be ready to take on their own project after having successfully defended the final thesis. Over the years of the PhD education, much knowledge about how science in the thesis subject works and about Academia in general is transferred to the student from the supervisor and from others in the academic environment.

Some of this knowledge is conveyed explicitly through courses, but much is more tacitly learnt by example, in particular from the supervisor or supervisors but also from others. This happens, for instance, through seminars and 'journal clubs' where science in the field is presented, discussed and critiqued. In the terminology of Kuhn and Lakatos, this is how the theoretical structures and accepted tools of the paradigm or research programme are learnt. What is an interesting research question to address and why? How do you properly design an experiment or perform observations to investigate it? Should the approach be strictly hypothesis testing or is some brand of empiricism permitted or even encouraged?

How do you analyse and present your results so that they are not only accepted by colleagues but also of interest to them? Or, in Ziman's terminology, this is how you learn what counts as scientific evidence in your field of research, evidence that can form the basis of a 'rational opinion' about some aspect of the world.

In the past we have frequently been involved in courses at the PhD level, at both introductory and later levels. If we asked students in the introductory course whether they considered themselves to be scientists, most said no. If we asked students in the third year of studies, most (but not all) said yes. When probed, those who said yes had already published the first chapter or chapters of the thesis, and, in some cases, had as a result been asked to act as peer reviewers for the scientific manuscripts of others. They might also have been to scientific conferences where they presented their results and discussed them with others. They felt that by now they belonged to a community of scientific colleagues – in other words, they started to see themselves as competent researchers and to be seen as such by others.

10.1.2 The Academic Career

Authors of scientific publications often have an address at a university or other academic institution, but not always. Sometimes the address is a company, a research institution or a governmental agency – employing scientists (often with a PhD, but they could also have a lower degree) pursuing a career outside of Academia. Medicine is another area with its own research traditions (varying among countries) where medical doctors typically have a focus on profession rather than research, but where some may choose to pursue scientific research and publish articles using hospital or other healthcare addresses. Similar situations apply in other professional fields with extensive and advanced paths of education. On rare occasions published authors or (more often) co-authors of scientific publications may lack an institution address altogether; the so-called amateurs who publish under their home address. This can happen if a person has built a reputation as a knowledgeable expert on a specific topic, such as astronomical observations or the natural history of some group of organisms, so that their contributions are valued by academic researchers even though they may lack formal scientific training. Thus, they are in a sense viewed as competent researchers, but only as long as they stay within their narrow topic.

Returning to the academic career, the first step after finishing PhD for a prospective scientist is the 'postdoc' stage, initiating a period of transition to what will hopefully eventually be a tenured position as a university teacher. The postdoc period often involves a stint at a university or research institution in another country. Such positions can be funded from grant agencies in the home country of the PhD, aiming to further train their future scientists as well as hoping that they will eventually bring back home useful new expertise. The lucky PhD who lands such a grant will have considerable agency in choosing their academic hosts as well as in designing the research project. Postdoc positions can also be announced by the host institutions when scientists have funding to employ persons to work in their existing projects. In such cases, the degree of independence in pursuing research can vary a lot, but when hosts value the input and complementary skills of their postdocs this is also a reciprocally rewarding situation furthering the careers of both.

Next in the academic career follows a somewhat nebulous stage where the PhD in most cases still lacks the merits (mainly in terms of scientific publications) to be competitive for a tenured university position. They may go on to another postdoc or similarly be temporarily employed as a researcher in a project funded either by a more senior scientist or even by the junior PhD themselves if they manage to get a grant for a project of their own. The hope is, however, to obtain a tenure-track position (terminology varies from country to country; they can, for instance, be called 'assistant professor' in the US and 'lecturer' in commonwealth countries). Promotion from this position to a tenured position ('associate professor' or 'senior lecturer') is generally not guaranteed but contingent on having achieved certain scientific and pedagogical criteria set out by the university. Failing to achieve tenure is a major blow to an academic career. Finally, the highest academic level is professor, and in some systems (such as the Swedish) this level can be reached either through promotion from a tenured position or when a position as professor is announced as such. Some universities make a strong distinction between these two paths to professorship, some do not.

It should be noted that at universities all academic positions in the tenure track are positions as university teachers, reflecting the expectation that scientists should take part in undergraduate education as well as do research and supervise PhD students. For many scientists, administrative duties of various kinds can also take up a large proportion of their time, by choice or not. Finally, many scientists take on positions of leadership and/or sit on department and faculty boards, a reflection of the fact that strategic decisions at universities about education, research and outreach should ideally be collegial and strongly influenced by active scientists.

10.1.3 Equal Opportunities in Science

In an ideal world, only talent and dedication should determine who becomes a successful scientist, one who is recognised as a 'competent researcher' and thus can influence the prevailing consensus in their chosen field of research.

Unfortunately, the world is not ideal. For one thing, in most Western countries the percentage of women drops with each step in the academic career. In many subjects (particularly the STEM disciplines of science, technology, engineering and mathematics) there can be more women than men at the undergraduate level, more equal numbers at the PhD student level, and then further reductions in the percentage of women at each career step until the professor level, which can often be heavily dominated by men.

There have been various attempts to explain this 'leaky pipeline' of women disappearing from science [47], with different emphasis on actual gender bias versus the active choices of women (and men) themselves. Notably there are many cracks in the 'pipeline'; gender differences can be found throughout the academic career. Kathleen Grogan has shown how the cracks can begin with women being less likely to be employed as PhD students or post docs in labs with male principal investigators (which is the majority) and continue with them being less likely to get excellent recommendation letters and having less chance of being published, getting grants and awards [48]. According to this summary, women also submit fewer grant proposals, one example of a 'crack' which can be the result of an interaction between gender bias and active choices. Another is the choice to go on a postdoc abroad, or to accept the general uncertainty of the academic career before tenure, which could conceivably present more of a challenge for women – given the prevailing gender roles. Whatever the cause, every step away from equal opportunities in science means that there will be unrealised potential in Academia; we will miss out on many competent researchers, some of whom potentially would have risen to excellence in their fields.

Bias in the scientific world similarly impacts in particular social and ethnic groups. Even in a country like Sweden with its free university tuition and comparatively generous student loans, students who have parents with an academic background are much more likely to pursue higher studies and eventually go into Academia. Students from ethnic minorities can have a harder time following an academic career as well. What is also obvious is that despite science in many countries being a very international endeavour, with research groups often consisting of many nationalities, this does not mean that opportunities are equal for scientists from different countries. English has emerged as the common language of science, meaning that countries where English is not learnt from an early age have a distinct handicap. Many such countries are faced with a dilemma where, on one hand, they want their scientists to have access to journals where they can publish in their own language, but on the other hand, such publications have much less chance of having an international impact. For one thing, key journal databases such as Web of Science (Chapter 8) index very few non-English journals, meaning that the publications will not be found through a literature search without extra effort.

In an attempt to lessen the impact of bias, scientific journals have increasingly turned to anonymous reviews of manuscripts. This does not work completely, because even without author names and affiliations, the research group in question can sometimes be obvious to a reviewer familiar with the field of research. However, it does prevent some of the bias (which can be subconscious from the side of the reviewer) arising from author gender, ethnicity and nationality. It will also lessen the positive bias that can easily arise from seeing a well-known authority in the field among the author names. The latter may be less of a problem than negative bias if we assume that such a scientist is an authority because they have made valuable contributions in the past, increasing the probability that the present results are also correct and of scientific interest. It is nevertheless a step away from an objective scrutiny of the manuscript and can allow inferior studies to slip through the peer review process.

10.1.4 The Invisible College

We have emphasised the importance of the scientific institutions and the role they play in maintaining the web of trust. However, the relationship between the scientist and these institutions is not necessarily without friction. To a large extent, this stems from a potential misalignment of goals and objectives. Scientists want to pursue their research questions and want the institutions to provide the necessary means to do so. This should obviously also be in the interest of the institutions, but they may also have interests of their own that are not always aligned with those of the scientists. Academic journals may want to make a profit from their publications. Universities may want to climb on national or international rankings, attract students or be visible in society, for example, or they may want to attract investors. Funding agencies may want to primarily fund research with high visibility to strengthen their own trademark. Or worse, they may want researchers to promote an agenda or products that they are associated with. Governments, who often provide the majority of research funding, may want to direct what types of research and education universities should provide society with and can sometimes have political agendas that extend beyond finding out the truth.

It is not unusual for universities and other academic institutions to expect a certain amount of loyalty from the researchers they employ or support. This can take many forms. From simply demanding that scientists highlight the institution in publications and presentations to expecting researchers to participate in events mainly aimed at promoting the institution or its agenda. The former is typically not problematic, and here the interests are often aligned. Researchers, may for example, have an interest in displaying funding information in their publications and presentations, partly for transparency and partly for self-promotion (to show that they have been deemed worthy of such a grant). It is also in the interest of the scientist to show an affiliation with a respected university. But beyond this, interests may not be fully aligned.

Universities have a long history of collegial governance, where many decisions are made by committees and boards with representation from the faculty as well as from students. This is to various extents upheld today but is also mixed with more management-orientated governance styles, where economic interests and political control may be more prominent and where an expectation of employee loyalty is more pronounced.

Most scientists actually do not see themselves as loyal to their employer, the way that employees of a commercial company would. This is not unique among public servants. A doctor, for example, should be loyal to their patients, not the hospital, and a teacher should be loyal to their pupils, not the school. But who does the scientist feel loyalty towards? It is not their academic department, not even their immediate colleagues at the department, and it is certainly not the university or the funding agency. No, the group of people they feel loyalty to is the collection of researchers working in the same field, exploring similar questions, and who they regularly meet at scientific conferences. John Ziman termed this the 'invisible college' [20], as it is a loosely connected group of researchers around the world, working in different academic environments but striving towards consensus on similar questions. The invisible college extends beyond the borders of even collegial governance at the university, connecting researchers from different parts of the world in an amorphous network. These people may well be fierce competitors, even enemies at times, but nevertheless, this is where the sense of loyalty is directed. The most prominent example of this network in action is the system of peer review that was described in Chapter 8. All researchers are dependent on the evaluation of other colleagues in the invisible college. They make up the 'competent researchers' within their particular field of science and are thus made up of the people participating in a shared research programme. These are the people whose opinion a researcher wants to affect and whose acclaim they seek. This is where their academic loyalty lies.

It is an important point to make since it is not necessarily well understood by the management of the academic institutions, who may expect or even demand loyalty of the people they employ or support. Still, if scientists were to truly show too much loyalty to these institutions, the web of trust would be in peril. It is the invisible college that is weaving the fabric of this web, while the role of the institutions is to protect its integrity and make the weaving possible.

10.2 Academic Freedom

The concept of *academic freedom* can be somewhat elusive,[1] even though such freedom is actually protected by international laws and, in many countries, also by national laws and constitutions. Academic freedom encompasses research, higher education and communication of science. When it comes to research, academic freedom has four components: the rights to freely choose research *questions* and research *methods*, the right to freely search and take part of *information*, and the right to freely choose *information channels* for research communication. Regarding education, academic freedom means that teachers have the right to base course content only on scientific knowledge without being constrained by other interests, and students have the right to freely choose subjects in order to learn about the world.

Naturally, total academic freedom can never be realised, because resources for research are not infinite and scientists typically belong to a collective and are constrained by the collective's choices. Publication channels can also be limited, in particular due to competition among scientists. The choices of students are constrained by background, resources and competition. However, it is important that such limitations are not intentionally put up for reasons in conflict with the scientific aims for research and higher education. In this, academic freedom means having the right (if not for individuals, at least for the collective) to reject loyalty demands stemming from national, political, ideological, religious or commercial interests, so that what *can* be done within the unavoidable constraints also is *allowed* to be done.

Why should there be academic freedom and why is academic freedom at least to some extent even protected in law? The main reason is to protect the right of citizens to obtain knowledge that is derived independently from other considerations, as objectively as possible. For a functioning democracy, citizens must be able to put some trust in the knowledge disseminated from scientists and universities, knowing that they are at least not intentionally misled.

---

[1] For our treatment of the concept, we follow a document produced by the Human Rights Committee of Sweden's Scientific and Literary Academies in 2025: 'Uttalande om akademisk frihet' (in Swedish).

10.3 Funding

One aspect of science that those outside of Academia may not fully appreciate is the need for funding. Early in a scientist's career funding can be crucial for the employment itself, perhaps as a postdoctoral researcher or as a non-tenured assistant professor, and is also needed for the actual research costs, travel costs and so on. Once the goal of a permanent employment has been achieved, the salary itself and many other costs coupled with the position are typically paid for by the university, but there is most often still a great need for funding to do research. Some scientists, such as theoreticians, can perhaps get by with just a computer, but for others both experimental and observational research can be quite costly.

Funding can be received after application to various grant agencies, in competition with other researchers answering the same call. We won't go into details of such agencies here, since there is much variation among countries and diversity within countries. Suffice to say that there are both national and international funding agencies, both governmental and private. In the competition to receive the funds, both the merits of the scientist (especially in terms of publications; see Chapter 8) and the proposed project matters, but to varying degrees depending on the agency and the specific call. Notably, the grading and ranking of grant proposals are almost always made by other scientists, even if the final decision may be up to the funding agency. The scientist peers will do their best to judge the proposed project and the merits of the applicant according to the criteria set out in the call, providing another key part of the web of trust.

Calls can be very broad and intended for funding basic 'curiosity-driven' research. In such cases the scientific merits of the applicants are very important for success, but also that the project seems feasible as well as shows novelty and has the potential to move the research field forward. Calls can also be more specific, and this is typically the case when there is an applied aspect to the research. It could, for instance, be that the government has decided to put aside research funding for a specific purpose which might help the industry (including agriculture, fishing and forestry) or help the country to achieve other goals such as environmental protection, clean water availability or public health. For such calls the proposed project is proportionately more important for success, in particular that it believably addresses a real need for research, as identified by 'stakeholders' (agencies, organisations and companies invested in the field).

Funding availability for basic vs. more applied research can be a matter of much debate. It is not surprising that politicians might want to direct governmental funding towards specific goals in order to make what they see as the best use of tax-payer's money, or that private funding agencies may have similar concerns about being associated with some applied research goal. However, there may come a point where too little funding is available for science to go in new and unforeseen directions, a point when the academic freedom of scientists to choose their projects is threatened. It might seem something of a luxury for scientists to have this freedom, if they are dependent on external funding. However, if you consider just about any technical breakthrough for humanity, be it the smartphone, the internet, antibiotics or vaccination, they would not have happened without a long history of scientists following their curiosity. We would argue that it is for such reasons essential that a great deal of funding continues to be available for basic research and that such open funding is guarded against the threat of an ever-increasing share of more directed funding.

How to apply for grants is almost a science in itself, and we will not cover this subject in depth here. It should be said, however, that it is essential to tailor the application to the funding agency and to the specific call in order to maximise the chances of success. It is also important to start the grant writing process well ahead of the deadline, since it can be a very time-consuming process. Actually, scientists spend a substantial fraction of their time writing applications for funding rather than doing research, another fact which has led to much debate, as well as to some attempts to help the situation. This could happen by, for instance, prolonging the grant periods so that there is no need to almost immediately start thinking about the next grant, or having a two-step process where only applicants surviving the initial screening need to write a full application, with an extended description of the intended project.

10.4 Ethics in Science

Research ethics is a complex and multifaceted subject which we can only touch upon here, focusing mainly on how ethical considerations interact with the 'web of trust' in science. When persons outside of the scientific community consider research ethics, odds are that they mainly think about various kinds of cheating and manipulation of scientific evidence, since examples of such blatant misconduct sometimes are highlighted by media. Any kind of cheating is also undoubtedly a grave danger for science, since it means that the public cannot wholly trust scientific findings, and scientists cannot trust each other. We will return to this important subject, but first note some other important facets of research ethics.

10.4.1 Research Subjects

Historically much of the debate around science ethics was concerned with the subjects for research, in particular in light of World War II with the advent of nuclear weapons. Is it ethical to even do research with the aim of producing new weaponry? Is it ethical if the weapons are only used for defence against aggressors? How can you know that this will be the case? It is up to the individual scientist to form an opinion on such questions, based on personal ethics.

10.4.2 Human and Animal Research

This is a related area of often difficult ethical considerations, where most countries have organised a system of ethical permits that have to be approved by some agency before a project can even start. The researcher has to motivate why it is important to involve humans or animals in the research and describe in detail how the project will be carried out. Considerations include how to ensure anonymising and/or safe-keeping of results concerning humans and minimising the suffering of animals. Many scientific journals demand that ethical permits be listed upon submission of the manuscript. Misconduct would include incorrect descriptions of the project in applications or claiming that ethical permits exist when in fact they do not or do not cover the experiment in question.

10.4.3 Plagiarising

This is in itself a broad category of misconduct, including using other people's text, ideas, grant applications or data and presenting them as your own, but extending also to, for example, citation techniques that do not properly give credit to sources in scientific publications. Like most forms of scientific misconduct, it can be understood as driven by the strong need for a steady output of strong publications in order to succeed as a scientist, but this is an explanation and not an excuse. The relevance for the web of trust is mainly that any form of plagiarising erodes the trust among scientists and with it the potential for fruitful collaborations and mutual respect. The increasing use of AI tools in science has exacerbated the problem, even to the extent that AI in extreme cases is used to generate entirely fake publications out of thin air, with authorship being sold. If this becomes a widely used practice, trust in science itself will be under severe threat. At the time of writing, discussions have just started on how to effectively detect and stop this form of misconduct.

10.4.4 Publication Misconduct

There are several other forms of potential misconduct involving the publication process. One such is when results are reported to the public before peer review, in particular if it later turns out that the results were not trustworthy in the first place (see also Chapter 8 regarding 'pre-prints'). The other side of the coin is that scientists may feel that the results are so important to report that they cannot wait for review, for instance, in matters involving environmental or health hazards. Misconduct applies particularly when this excuse is used to present sensational results in order to further the careers of the scientists involved, with little regard of whether the results will actually stand in the end. Such behaviour is not helpful towards public trust in science. The same could be said regarding publishing biased reviews and opinion papers, where the decision on what scientific findings to include and which conclusions to make from them are driven more by personal or political agendas than by a desire to report current knowledge as objectively as possible.

When new results are published in peer-reviewed journals, there are some other behaviours that may be considered unethical. One example, also driven by the need for increasing the number of publications, is when the results are divided up into small pieces (one result per publication in the extreme case), or new manuscripts are created from old ones without adding new results or ideas of any substantial value. Whether such behaviour is truly unethical or not would depend on the actual intent as well as on the precise details of the particular case.

Authorship is another area of possible unethical behaviour. It might involve listing somebody as a co-author who did not actually substantially contribute to the scientific content of the manuscript, or conversely accepting co-authorship yourself without having made every effort to understand and validate the methods and results so that you can stand behind them.

10.4.5 Conflicts of Interest

Unethical behaviour can arise from conflicts of interest that introduce a bias in the interpretation and/or communication of scientific results, thus deviating from the aim in science of presenting the results as objectively as possible in order to aid in the 'search for a consensus of rational opinion' regarding how the world actually is.

The most obvious source of such bias is when the research is wholly or partially paid for by companies with a clear interest in the nature of the results, most blatantly when it comes to, for example, the tobacco, oil or medical industry, but also (and sometimes less obviously) in areas such as food, fishery and forestry. There is a paradoxical situation present at many universities, where collaborations with the industry are strongly encouraged as a source of more funding as well as part of strivings for outreach to the community, at the same time introducing a risk for bias that may at times end up in borderline unethical behaviour.

Funding from companies creates clear-cut possibilities for conflicts of interest, but similar situations can actually arise when funding is from governmental agencies with clear aims such as conservation of biodiversity or efficient use of natural resources (more about this in Section 10.5). Most scientific journals demand that any conflicts of interest are stated in the manuscript, but this is no guarantee that more subtle sources of bias are revealed.

10.4.6 Cheating and Manipulation of Results

Coming back to this central area of unethical behaviour, 'cheating' can take many forms. It may, in the extreme case, amount to a complete fabrication of data, but more commonly it involves some form of manipulation of the data. One example is when only data points or results that support a particular hypothesis are reported, ignoring data that go counter to it. A lack of documentation of the work often goes hand-in-hand with this form of manipulation. Another example is a deliberate over-interpretation of the data so that it better supports the hypothesis, perhaps including selection of which results to highlight more in the manuscript, which image to show as the 'typical' one or even 'improving' the image so that it better shows what you want it to show. A third example is distorting the work of other scientists so that it better fits your hypothesis, or minimising the results from others when they are in conflict with your own results, or if you want to claim an idea as your own.

It is important to realise that with the exception of actual fabrication of data none of these situations are completely clear-cut. There is often a grey zone where it can be debated whether a behaviour is actually unethical or just part of streamlining the manuscript to make the take-home message clearer. Again, whether real misconduct is present comes back to the intent and to the details of what was done.

Finally, we would like to note that manipulation of data is not only detrimental to the progress of science, in that it may lead to other scientists spending time and funding to pursue avenues of research based on findings that were less strongly supported than it seemed from publications. Cheaters are also fooling themselves. We are inclined to believe that most cases of manipulations of results arise because the scientist really believes in their own results and 'only' wishes to make them clearer and more publishable. However, there is a strong risk that this only leads to the scientist clinging to a hypothesis that should in fact have been abandoned. Cheating and manipulation lead nowhere – even if you do it for selfish reasons to momentarily further your career, it leaves you with no strong foundation for your own future work.

10.5 Science Outside of the Academy

> The worst of scholars are those who visit princes, and the best of princes are those who visit scholars[2]

Science has an important role to play also outside of the academy, but this is a potentially troubled relationship. Science is dependent on other parts of society for funding and other kinds of support, yet at the same time it is important that the work that scientists do remains independent from outside influence. We have already explained the delicate balance that needs to be maintained with regard to unbiased research funding from the point of view of the scientist. Likewise, while governments and companies depend on access to unbiased knowledge in order to make informed policy decisions and to make meaningful plans for future development, they may still find it tempting to instead seek answers that align with their own agendas. The independence of science is as important for the decision-makers as it is for the scientists, even if the extent to which this is truly understood and embraced may vary. Moreover, even with the best of intentions, seeking out scientific knowledge to make informed policy decisions is not necessarily an easy task, especially for someone outside of the academic system. Scientific findings are often difficult to penetrate and evaluate for non-experts, so how can society gain access to and truly benefit from such findings?

10.5.1 The Challenge of the Consensus

We have written extensively about the search for consensus among scientists in this book. When a person outside of academia turns to science for the answer to a particular question, what they really ask for – whether they realise it or not – is the current scientific consensus on this question. Unfortunately, this

---

[2] Jalal al-Din Rumi (1207–1273), in 'Discourses of Rumi'. A prescient quote on the relationship between academia and political power. It is important for scholars to remain independent from political power, but equally important for people in power – and the society at large – to seek out the knowledge of scholars, and to take it seriously.

consensus is not always easily grasped, especially for contentious questions or in very active research fields. Moreover, the consensus is not static; it changes as researchers continue to challenge and improve it. The current consensus is therefore at best a fleeting snapshot and is not always easy to articulate, even for a scientist (for an illustrative example, see Section 9.4 on combining evidence).

Science typically reaches the general public only in bits and pieces and almost always out of context. The rise of social media as the primary carrier of news and other information has not made the situation easier. Social media allows information to travel fast but also tends to emphasise controversy over consensus. Academics can contribute to muddying the waters, too. Scientists are often asked to opine on topics in interviews and on TV-sofas. In itself this is positive, but most scientists are rather reluctant to play this role. We are so mindful of the tentativeness of scientific findings that we may end up being overly cautious. Journalists, on the other hand, tend to prefer clear and uncomplicated messages, and when a scientist does step up to take this role, they will be welcomed and increasingly sought after. For such a scientist, who perhaps thrives in the limelight, it can be tempting to increasingly play the role of 'expert', also on topics that are out of their competence. It stands to reason, then, that for someone outside academia, grasping the scientific consensus on a given topic is a formidable task.

It may seem obvious that if someone from outside of academia really wants to find a scientific answer to a question, they should consult the actual scientific literature on the topic. Unfortunately, as we saw in Chapter 8, more often than not this literature is locked behind publisher paywalls and accessible only via university libraries. Even if one manages to access the articles, scientific papers are not necessarily easy to penetrate, and the sheer number of published articles makes the task even more daunting. For an inexperienced questioner it will be all too easy to skim the scientific literature until they find an article addressing the question at hand, and then be content with having found the answer (or a 'source' for their claim). But a consensus must necessarily involve the work of more than one researcher. Since science is unavoidably a trial-and-error process, the chances of any particular published result being wrong are also much higher than the chances of the scientific consensus being wrong. Hence, it is not enough to be content with finding one or a few papers addressing the question. You need much more than that to be able to gauge the scientific consensus.

10.5.2 Evaluating the Current Scientific Consensus

As we saw in Chapter 9, different types of scientific publications provide different types of data and thus different input for judging where the current scientific consensus stands on a topic of interest.

An active researcher will over time develop a good 'feel' for the consensus in their particular field of research. For this reason, they are sometimes called upon by a scientific journal to write a scientific review on a given topic within their expertise. Such reviews summarise the current state of the art on a topic and also typically point out areas that require more attention. Such reviews can be very valuable, not least for someone who is new to a field. They contain both a survey of the current consensus and a large number of references to relevant original research articles to investigate further. Still, it is important to not stop there. If the task is to really evaluate the veracity of a claim or to provide an accurate description of the current consensus, such a review should be the starting point, not the end point. For one thing, it is rare that the review addresses exactly the question you have in mind. Moreover, the review is filtered through the mind of one or a few researchers, and it is not necessarily the case that the view of the topic that emerges is entirely free of personal opinion. If the review is more than a few years old, it is also quite possible that new research has emerged that would require its conclusions to be adjusted. It is therefore important to also consult original research articles, that is, studies containing original data in some form.

Such studies can come in different forms, and as we saw in Chapter 9, it is often necessary to puzzle together the answer by combining information from several types of studies (experimental, longitudinal, structured observational studies, meta-analyses, etc.). It is worth reminding us here how both the observation-driven (Chapter 2) and the hypothesis-driven (Chapters 3 and 4) theories of science ended up emphasising the importance of gathering multiple kinds of evidence in order to strengthen a hypothesis. When investigating the veracity of a claim, you can easily see why this makes sense. For example, longitudinal studies in medicine or nutrition can follow thousands or tens of thousands of subjects over time, and can through statistical analysis unravel correlations with, for example, mortality that otherwise would be hard to detect. On the other hand, they offer little insight into actual causality. Experimental studies, although having far fewer test subjects, can do just that. They are in a much better position to unravel if and why, for example, eating a lot of processed meat causes excess mortality. Such experiments may have to be done on rats but still contribute important information when combined with other studies.

When scientific results are evaluated and communicated, it is hence important to judge which type of study you are dealing with and to what extent it can address causality. Say that the research question is whether the mandatory wearing of face masks helps stop the spread of a pandemic. One approach is to use existing data to perform an observational study: When comparing countries or regions, is there a correlation between the reported number of cases and the stringency of face mask mandates or (better) the percentage of people actually wearing them? As we saw in Chapter 9, the results of such a study might provide some answers, but there is a high risk of confounding variables since the areas studied will differ in many other ways. A better set-up would be to make a number of pairwise comparisons of nearby regions differing in mandates in a pseudo-experiment, but confounding factors would still be a problem. If at all possible, it is highly preferable to get closer to a real experiment, where instead of looking back at existing data you do the manipulation yourself. In a country where mask wearing is not mandated, you might be able to enrol a few villages or schools to somehow encourage the practice. As part of the study, you assess the actual mask wearing across the study sites (Was the manipulation successful?) and later the number of cases of the disease reported or perhaps the number of hospitalisations or deaths due to the disease. If the villages or schools can be assigned randomly this would go much further towards demonstrating causality – or indeed the lack of a correlation. Finally, laboratory experiments studying the effect of face masks on dispersal of the contagion can be done in a much more controlled way. Such studies can thus in a superior way address the *potential* of masks for reducing the spread of the pandemic, but may on the other hand, by themselves say rather little regarding the complex situation in the real world.

10.5.3 Making Science Actionable

It should be clear that the investigative work on scientific consensus requires a substantial level of expertise, if not with the actual topic at hand, then at least with finding, reading and evaluating scientific literature in general. In addition to navigating the maze of scientific studies outlined in Section 10.5.2, an investigator will, for example, also need to know the difference between peer-reviewed articles and preprints, and be able to recognise 'predatory journals' where peer review is mostly lip service. It is thus a highly qualified work that typically falls to scientifically trained professionals who are not necessarily active scientists but who work as scientific consultants. Their work is a tremendously important part of the web of trust, since it can have direct influence on far-reaching societal decisions and thus has to provide a fair and accurate representation of the scientific evidence. The typical output of these professionals is not research articles in scientific journals but a different class of publications called reports. The target audience of the report is different from a scientific publication, and so is its purpose. While the scientific article is targeting other researchers with the goal of affecting the scientific consensus, the goal of the report is to communicate this consensus to people that do not necessarily have scientific training, but that may well hold decision-making positions in the society. These decision-makers could be the leadership of a private company or organisation but very often they are the political leaders and public servants of various governmental or intergovernmental institutions.

Reports are not peer reviewed in the way that scientific articles are, as they do not contribute new scientific knowledge. They should instead provide the necessary scientific background to the topics that our policymakers have on their table. These topics can range from global consequences of climate change to more localised effects, such as the effects of new road construction on a local frog species. Regardless of the scope, it is important that the report is actionable and not too vague. This can be a challenge, since there is often a level of uncertainty involved in scientific conclusions. In a scientific article it may be permissible (even advisable) to use expressions such as 'may' or 'could' to signal such uncertainty, since, as we have seen, acknowledging the fallibility of scientific findings is an essential part of the scientific mindset. In a report, however, such vagueness will only serve as an excuse for the policy-makers to disregard scientific knowledge in favour of other pressing agendas.

To meet this challenge, there has been an increasing drive to develop systematic guidelines that explicitly address levels of uncertainty. After all, some scientific knowledge is virtually certain, while other findings are more tentative, and it is important that this is communicated to the policymakers. One example of an effort to address this issue is the GRADE (Grading of Recommendations Assessment, Development and Evaluation) criteria, initially developed to improve the quality of guidelines for the health sector. Similarly, the urgency of communicating the complex scientific knowledge on climate change prompted the IPCC (The Intergovernmental Panel on Climate Change) to develop a standardised system of communicating scientific knowledge to policymakers. These efforts have since influenced the way reports are written on many other topics.

In addition to assessing the risk, impact or severity of an outcome, there is also an assessment of scientific confidence – essentially to what extent scientists in the field agree on a given conclusion. This is especially important for questions that may be politically controversial, where scientific disagreement can be weaponised in order to weaken scientific arguments. It is not uncommon for political interests to promote one or a few scientific dissenters in order to argue that there is in fact no scientific consensus. We can see here how the IPCC, in response to this, has developed an explicit approach to estimate the strength of the scientific 'consensus among all competent researchers', and present it in a way that is comprehensible also for someone outside this group.

These types of graded assessments of the scientific consensus are today routinely used in scientific reports directed at policymakers. The benefits are clear: it provides the target audience with direct and transparent information on the strength of the evidence, its level of certainty and the severity of the impact. This is useful, as it allows policymakers to target the outcome with the most severe effect and the highest confidence estimate first. Other criteria can be included, too, such as the feasibility of a suggestion or its acceptability to various stakeholders. By using such deliberate criteria, the authors of a report will not only make scientific results available to an audience outside academia, but they will also make sure that the conclusions are operational. Stakeholders are provided with reliable arguments for action.

11
Epilogue

Science is a web of trust.

In Chapter 1, we noted how one goal of science is to make subjective observations more objective, and how this typically requires some active work on the part of the observer. We further noted that it is virtually impossible to make theory-free observations. All observations require an interpretive framework and are dependent on pre-existing assumptions about the world. Sometimes this theory-dependence is obvious, such as when reading the output of complex machinery; sometimes it is more subtle, but even the simplest of observations require pre-existing theory to make any sense. An important consequence of this is that observations themselves are fallible.

In the chapters that followed, we made the somewhat paradoxical observation that even if this fallibility of observations has been integral to the success of science (because it allows our understanding of the world to progress), it has turned out to be a stumbling block for virtually all attempts to formulate a solid method-based philosophy of science. This was most plainly true for empiricism, with its emphasis on unbiased observations, but it also turned out to be a problem for the more theory-based philosophies that followed. Even if these approaches explicitly embraced theory-dependence, it turned out that it made the interpretation, for example, of an experimental outcome less conclusive than one would wish. All these approaches were in need of a framework to navigate the complex maze of assumptions and provide guidelines for the difficult decisions involved with the interpretation of observations.

We then turned to some attempts to provide just this, in the form of paradigms and research programmes. These are theoretical constructs that provide the interpretive framework that can help researchers make sound methodological decisions, and focus on the important questions at hand. Moreover, such interpretive frameworks assure that researchers at any given time and field share the same fundamental assumptions, providing cohesiveness to their work and preventing fragmentation. Still, for all their usefulness in describing how science works, paradigms and research programmes do not provide a clear definition of science. They do, however, pave the way for a social definition of science that focuses on the search for consensus of rational opinion among researchers working within a paradigm or research programme. Being method-agnostic, an attractive property of this definition is that it allows scientific methods and standards to evolve. Another consequence is that it becomes more inclusive, in that it can accommodate past, current and future research using other methodological approaches than what is currently accepted as the norm within the natural sciences. We further argued that the 'line of demarcation' between science and other means of gaining knowledge is more of a grey zone than a clear-cut line.

In the end, we had to conclude that the question of what science is has no single answer, but this sounds more pessimistic than it truly is. The reason why we cannot find a single answer to the question is because it is in fact several questions disguised as one. Explaining *what* science is, is not the same as explaining *how* it has been done, is done or *should* be done. Moreover, science is a moving target, in constant development, and any attempt to tie it down to a definition needs to acknowledge this. We ended this part of the book with an elaboration of what it means to be *scientific*. Although clearly an important aspect of understanding science, it also showed how such descriptions cannot work as a definition of science, since it is quite possible for a scientist to behave unscientifically, and for a non-scientist to have a scientific approach to their work. This also highlights another pattern that should emerge from the first part of this book: throughout these chapters, we could see a gradual shift in focus from the actions of individual scientists to the larger theoretical, or even societal, structures in which the scientists work. Science is not the behaviour of individual scientists but an emerging property of their collective work. Individual scientific findings do not become part of science until they are made public and subjected to scrutiny by other researchers. This process of uncovering, publishing and discussing scientific findings is also heavily dependent on support structures such as public education, universities, libraries, funding agencies and publishers. These structures must all work towards ensuring the integrity of the scientific work; they are all parts of the web of trust. They are an integral part of science, as scientific work would not be possible without them. For this reason, we believe it is misguided to attempt an understanding of what science is without acknowledging this. In an important sense, we could even say that science *is* the web of trust.

11 Epilogue 151

In the last few chapters, we took a deeper look at this web of trust as it manifests itself today and suggested that the 'social definition' of science proposed in Chapter 6 provides a useful framework to make sense of its intricate workings. First, we examined what it means today to make scientific knowledge public by describing in some detail how scientific publishing currently works. Then, we examined 'data' as a common currency for arguing a rational opinion. In particular, we discussed if and how statistics are used to interpret such data, and showed how deeply intertwined statistical considerations are with the philosophy of science. In Chapter 10, we looked at what a 'competent researcher' means today and the role of the academic institutions in assuring and maintaining such competence. Here, we also pointed out some problematic aspects of academia that unless checked can threaten the integrity of the web of trust. We ended the chapter with an overview of the relationship between science and the rest of society and the important role that scientifically trained individuals working outside of academia have in strengthening this relationship.

The answer to *what* science is can thus help us to understand also *how* it is done. The final question of how science *should* be done – typically the primary focus of philosophy of science – is still a work in progress. It is not that we haven't made any headway with this question, but we must acknowledge that we still may not have quite figured out the best way to learn about the world. We suppose some of you may feel a bit disappointed by this. But on the other hand, is there not a certain excitement in the realisation that this remains an open-ended quest, still in need of creative work and refinement? If, as Ziman has suggested, doing science is like making a map of a country you cannot visit, the methods of charting out this territory are continually being refined. We have come a long way, but much of the road ahead is still uncharted.

The safeguarding mechanisms that make up the web of trust have emerged and grown organically as the scientific community has encountered and adapted to new challenges. They must also continue to evolve to meet current and upcoming challenges, such as those at the time of writing coming from the rapid rise of AI-based tools, and their continued success in doing so should not be taken for granted. The web of trust is under constant threat, sometimes directly and deliberately, because individuals or organisations can gain short-term benefit from undermining the trust people put in scientific results. Sometimes more indirectly, such as when governments, organisations or companies want to direct funding to where it suits their agenda, threatening academic freedom. Some threats and challenges come from within the scientific world itself. The various forms of scientific misconduct that we outlined in Chapter 10 are obvious examples. But publishing in journals with questionable quality control, exemplified by the 'predatory journals' mentioned in Chapter 8, can also erode the web of trust by making it harder to trust that a paper has undergone rigorous scrutiny before publication.

In order to protect the integrity of the web of trust, the scientific community must recognise both its importance and its fragility. Weaving the web of trust is a responsibility of the whole academic world, but also of society in general. It is safe to say that our modern society would not exist without science, and that continuing societal development is highly dependent on the integrity of scientific findings. If science is to work properly, also outside the corridors of the academic institutions, our societies need an educated population capable of making informed decisions rooted in scientific consensus. Thus, we must acknowledge that the web of trust extends beyond academia and reaches into the everyday lives of us all.

References

1. R. Blondlot, *'N' Rays: A Collection of Papers Communicated to the Academy of Sciences* (London: Longmans, Green, and Co., 1905).
2. F. Darwin, ed., *The Life and Letters of Charles Darwin, Including an Autobiographical Chapter* (London: John Murray, 1887).
3. C. G. Hempel, 'Studies in the logic of confirmation (I)'. *Mind*, **54**(213): (1945), 1–26.
4. W. L. Johanssen, *Elemente der Exakten Erblichkeitslehre* (Jena: Gustav Fischer, 1909).
5. D. Mayo, *Error and the Growth of Experimental Knowledge* (Chicago: University of Chicago Press, 1996).
6. C. Darwin, *The Origin of Species* (London: Penguin Books, 1859).
7. K. R. Popper, *The Logic of Scientific Discovery* (London: Routledge Classics, 1959).
8. K. R. Popper, *Conjectures and Refutations: The Growth of Scientific Knowledge* (New York: Basic Books, 1962).
9. K. R. Popper, 'Normal science and its dangers'. In I. Lakatos and A. Musgrave, eds., *Criticism and the Growth of Knowledge* (Cambridge: Cambridge University Press, 1970), pp. 51–58.
10. P. K. Feyerabend, *Against Method: Outline of an Anarchist Theory of Knowledge* (London: New Left Books, 1975).
11. C. Howson and P. Urbach, *The Bayesian Approach*, 3rd ed. (Chicago: Open Court, 2006).
12. M. Bayes and M. Price, 'An essay towards solving a problem in the doctrine of chances. By the late Rev. Mr. Bayes, F. R. S. Communicated by Mr. Price, in a letter to John Canton, A. M. F. R. S'. *Philosophical Transactions (1683–1775)*, **53**: (1763), 370–418.
13. J. Williamson, *In Defence of Objective Bayesianism* (Oxford: Oxford University Press, 2010).
14. J. M. Joyce, 'The development of subjective Bayesianism'. In D. M. Gabbay, J. Woods and A. Kanamori, eds., *Handbook of the History of Logic* (Boston: Elsevier, 2004), pp. 10–415.
15. D. Blackwell and L. Dubins, 'Merging of opinions with increasing information'. *The Annals of Mathematical Statistics*, **33**(3): (1962), 882–886.

16. C. Darwin, ed., *The Foundations of the Origin of Species: Two Essays Written in 1842 and 1844* (Cambridge: Cambridge University Press, 1909).
17. T. S. Kuhn, *The Structure of Scientific Revolutions*, 2nd ed., enlarged. (Chicago: University of Chicago Press, 1970).
18. I. Lakatos. *The Methodology of Scientific Research Programmes: Philosophical Papers*, J. Worrall and G. Currie, eds. (Cambridge: Cambridge University Press, 1978).
19. I. Lakatos, 'Falsification and the methodology of scientific research programmes'. In I. Lakatos and A. Musgrave, eds., *Criticism and the Growth of Knowledge* (Cambridge: Cambridge University Press, 1970), pp. 91–195.
20. J. Ziman, *Public Knowledge: An Essay Concerning the Social Dimension of Science* (Cambridge: Cambridge University Press, 1968).
21. J. Ziman, *Reliable Knowledge: An Exploration into the Grounds for Belief in Science* (Cambridge: Cambridge University Press, 1978).
22. J. Ziman, *Real Science: What It Is, and What It Means* (Cambridge: Cambridge University Press, 2000).
23. A. Cromer, *Uncommon Sense: The Heretical Nature of Science* (Oxford: Oxford University Press, 1995).
24. A. H. Maslow, 'A theory of human motivation'. *Psychological Review*, **50**(4): (1943), 370–396.
25. R. J. Kemkes and S. Akerman, 'Contending with the nature of climate change: Phenomenological interpretations from Northern Wisconsin'. *Emotion Space and Society*, **33**: (2019), e100614
26. J. Bronowski, *Science and Human Values* (New York: Julian Messner, Inc., 1956).
27. M. Baldwin, 'Scientific autonomy, public accountability, and the rise of "peer review" in the Cold War United States'. *Isis*, **109**(3): (2018), 538–558.
28. A. Fleming, 'On the antibacterial action of cultures of a *Penicillium*, with special reference to their use in the isolation of *B. influenzae*'. *British Journal of Experimental Pathology*, **10**(3): (1929), 226–236.
29. J. C. Stanley, 'Common class of pseudo-experiments'. *American Educational Research Journal*, **3**(2): (1966), 79–87.
30. T. Knuuttila and M. Boon, 'How do models give us knowledge? The case of Carnot's ideal heat engine'. *European Journal for Philosophy of Science*, **1**(3): (2011), 309–34.
31. T. Heger, A. Elliot-Graves, M. I. Kaiser, K. H. Morrow, W. Bausman, G.P. Dietl, et al., 'Looking beyond Popper: How philosophy can be relevant to ecology'. *Oikos*, **2025**: (2025), e10994.
32. R. Levins, 'Strategy of model building in population biology'. *American Scientist*, **54**(4): (1966), 421–431.
33. J. Ridley, N. Kolm, R.P. Freckelton and M. J. G. Gage, 'An unexpected influence of widely used significance thresholds on the distribution of reported P-values'. *Journal of Evolutionary Biology*, **20**(3): (2007), 1082–1089.
34. K. Kovaka, 'Mate choice and null models'. *Philosophy of Science*, **87**(5): (2020), 1096–1106.
35. C. Darwin, *The Descent of Man, and Selection in Relation to Sex* (London: John Murray, 1871).

36. A. E. Derocher, M. Andersen and O. Wiig, 'Sexual dimorphism of polar bears'. *Journal of Mammalogy*, **86**(5): (2005), 895–901.
37. K. Ralls, 'Mammals in which females are larger than males'. *Quarterly Review of Biology*, **51**(2): (1976), 245–276.
38. K. J. Tombak, S. Hex and D. I. Rubenstein, 'New estimates indicate that males are not larger than females in most mammal species'. *Nature Communications*, **15**: (2024), e1872.
39. J. M. Plavcan., 'Sexual dimorphism in primate evolution'. In C. Ruff, ed. *Yearbook of Physical Anthropology*, vol. 44., pp. 25–53 (Hoboken NJ: John Wiley & Sons, 2001).
40. A. G. McElligott, M. P. Gammell, H. C. Harty, D. R. Paini, D. T. Murphy, J. T. Walsh, et al., 'Sexual size dimorphism in fallow deer (*Dama dama*): Do larger, heavier males gain greater mating success?' *Behavioral Ecology and Sociobiology*, **49**(4): (2001), 266–272.
41. R. A. Fisher, *The Genetical Theory of Natural Selection* (Oxford: Clarendon Press, 1930).
42. R. Lande, 'Sexual dimorphism, sexual selection, and adaptation in polygenic characters'. *Evolution*, **34**(2): (1980), 292–305.
43. J. P. Reeve and D. J. Fairbairn, 'Predicting the evolution of sexual size dimorphism'. *Journal of Evolutionary Biology*, **14**(2): (2001), 244–254.
44. N. Korkman, 'Selection with regard to the sex difference of body weight in mice'. *Hereditas*, **43**(3–4): (1957), 665–678.
45. T. Janicke and S. Fromonteil, 'Sexual selection and sexual size dimorphism in animals'. *Biology Letters*, **17**: (2021), e20210251.
46. L. Winkler, R. P. Freckleton, T. Székely and T. Janicke, 'Pre-copulatory sexual selection predicts sexual size dimorphism: A meta-analysis of comparative studies'. *Ecology Letters*, **27**: (2024), e14515.
47. A. N. Pell, 'Fixing the leaky pipeline: Women scientists in academia'. *Journal of Animal Science*, **74**(11): (1996), 2843–2848.
48. K. E. Grogan, 'How the entire scientific community can confront gender bias in the workplace'. *Nature Ecology and Evolution*, **3**(1): (2019), 3–6.

Index

academic freedom, 137
ad hoc hypothesis, 39
AI, 26, 90–91, 140, 151
alternative medicine, 88
analytic philosophy, 81
antirealism, 2–3, 19
Aristotle, 16, 25

background knowledge, 44–45, 47, 93, 95
Bacon, Francis, 15
Baconian method, 16
Bayes, Thomas, 48
Bayes' theorem, 49
Bayesianism, 48–54
bibliometrics, 104–108
Big Data, 26
biological variation, 116–119

cohort studies, 112
Comte, Auguste, 18
conditional probability, 49
confounding factors, 124
confounding variables, 112
consensibility, 75
continental philosophy, 81
Copernican Revolution, 59
corroboration, 43, 45
COVID-19, 9

Darwin, Charles, 17, 30
Darwinian Revolution, 59
Darwinism, 60
data-driven science, 26
deduction, 15
demarcation, 34, 46, 69, 81–85

Duhem–Quine thesis, 42, 119
Dutch books, 51

empiricism, 14, 19, 26
ether, 6

fact, 5–9
falsifiability, 34–35
falsification, 31–34
Feyerabend, Paul, 46
Flores Man. *See Homo florensis*

gene, 23
Google Scholar, 103
GRADE criteria, 147
Grand Canyon, 22
Great Unconformity, 23

hard core, 63
hermeneutics, 83
H-index, 105
Homo florensis, 111
homology, 60
Hume, David, 20
Husserl, Edmund, 83
hybrid journals, 103
hypothesis, 9

impact factor, 106
incommensurability, 58–59
induction, 14, 19, 33, 114
    problem of, 20
inductivism, 14
invisible college, 136
IPCC, 147

Index

Kant, Immanuel, 82
Kuhn, Thomas, 57

Lakatos, Imre, 63
leaky pipeline, 134
logical positivism, 19

Mayo, Deborah, 27, 65
meta-analysis, 126
model organisms, 116

natural experiments, 112–113
natural theology, 60
negative heuristics, 64
new experimentalism, 27, 65–66
Newton, Isaac, 17
normal science, 57
N-rays, 7
null hypothesis, 119–121

objective Bayesians, 52
objectivity, 87
observation, 5, 22
Occam's razor, 14
open access, 99–100

paradigm, 56–63
parsimony, 14, 87
peer review, 73, 92, 97, 101, 141
penicillin, 111
phenomenology, 83
phlogiston, 6
Popper, Karl, 30–34, 37
positive heuristics, 64
positivism, 14, 18–19, 78
post hoc hypothesis, 39
posterior probability, 49
postmodernism, 84
predatory journals, 100
prediction, 9, 36, 119
pre-print, 92, 98
pre-science, 57

prior probability, 49
problem of the priors, 51–53
protective belt, 64
pseudo-experiments, 112
pseudoscience, 85
public knowledge, 71, 75
p-value, 121–123

quasi-experiments, 112
quasi-science, 85, 114

raven paradox, 20, 50
realism, 2–3, 19
repeatability, 86
replication crisis, 118
research ethics, 139
research programme, 63–65
retractions, 103

science-scepticism, 10
scientific consultants, 146
scientific models, 115–116
scientific progress, 18, 44
scientific revolution, 57
Scopus, 103
Shanghai ranking, 108
social definition of science, 75
social media, 4, 9, 100, 144
speciation, 70
species, 23
subjective Bayesians, 52

technology, 72
theory, scientific, 9
theory-dependence, 8, 21–25, 32, 42, 95, 114
transparency, 87

verifiability, 19

Web of Science, 103

Ziman, John, 70

For EU product safety concerns, contact us at Calle de José Abascal, 56–1°, 28003 Madrid, Spain or eugpsr@cambridge.org.

www.ingramcontent.com/pod-product-compliance
Ingram Content Group UK Ltd.
Pitfield, Milton Keynes, MK11 3LW, UK
UKHW020659180426
470123UK00009B/90